Advances in Anatomy Embryology and Cell Biology

Vol. 100

Joseph Altman   Shirley A. Bayer

The Development of the Rat Hypothalamus

With 125 Figures

Springer-Verlag
Berlin Heidelberg New York
London Paris Tokyo

Professor Joseph Altman, Ph.D.
Lilly Hall of Life Sciences, Purdue University
West Lafayette, IN 47907, USA

Professor Shirley A. Bayer, Ph.D.
Indiana University – Purdue University
1125 East 38th Street
Indianapolis, IN 46223, USA

ISBN-13:978-3-642-71303-3 e-ISBN-13:978-3-642-71301-9
DOI: 10.1007/978-3-642-71301-9

Library of Congress Cataloging-in-Publication Data
Altman, Joseph, 1925 –. The development of the rat hypothalamus. (Advances in anatomy, embryology, and cell biology; v. 100) Bibliography: p. Includes index. 1. Hypothalamus. 2. Rats – Development. 3. Mammals – Development. I. Bayer, Shirley A. (Shirley Ann), 1940 –. II. Title. III. Series: Advances in anatomy, embryology, and cell biology; v. 100.
QL938.H94A48 1986 599.32′33 86-10217

Softcover reprint of the hardcover 1st edition 1986

2121/3140-543210

Editorial

The present monograph is Vol. 100 of *Advances in Anatomy, Embryology and Cell Biology*. It marks a special occasion for looking back, but also into the future.

This monograph series was started in 1891 – almost 100 years ago – under the title *Ergebnisse der Anatomie und Entwicklungsgeschichte*. This title was retained until 1973, when it was changed to the present one. What has not been altered over the years is the scientific intention of the series. The purpose at its inception was the publication of reviews on topics in the field of morphology; this is still the policy behind the series, as it will be in the future.

Particular attention has been devoted to the link-up between morphology and the related disciplines in both scientific and clinical sectors. Overall, the 100 volumes reflect the tremendous progress made in the morphological sciences since 1891. It is interesting in this connection that the first 20 years of the series' existence were characterized by a predominance of topics concerned with macroscopic anatomy, while in the subsequent 20 years biogenetics and phylogeny were the main areas treated. Then, from about 1930 onward, the emphasis started to shift to micromorphology and in the 1960s to neuroanatomy. Nonetheless, there has never been any question of neglecting the other subject areas; cell biology and functional anatomy in particular have always been the subject of lively interest. The tremendous advances made in morphology also become apparent when we consider the methods used to process the results, especially since the introduction of electron microscopy, histochemistry and computer-aided quantitative procedures.

The first editor of this monograph series, F.R. Merkel, wrote prophetically in his preface (1891) that "so-called fine anatomy, which takes its methods from the exact sciences, physics and chemistry, and the finer anatomy of the central nervous system are the main items on the agenda." With regard to detailed research into gross anatomy, however, he said there was "the feeling of a sort of ending."

If we look back on these statements from our present position it becomes apparent that the points of emphasis addressed in 1891 are still in the foreground for the present and future: discussion of neurobiological questions and further advancement into the molecular sector of micromorphology, that is to say into the area where the medical sciences are most deeply rooted and where all branches of medical science meet. As in 1891, the universal significance of these two sectors means that they touch on many subareas of anatomy.

It can be inferred from the above that *Advances in Anatomy, Embryology and Cell Biology* is dedicated equally to tradition, to the present and to the future. The series has always been carefully adapted to the needs of the particular time, in close collaboration between publisher and editors. Originally only commissioned articles were published, and these were combined to make up quite large volumes, but since the number of people working in the field of morphology has increased so substantially the contributions have been published as separate volumes. This makes faster publication possible. In addition, the acceptance criteria of the series have been widened to include review-type articles in which the authors' own long-term research is summarized and the conclusions on specific points are presented. This feature is to be retained in the future. Another purpose fulfilled by *Advances in Anatomy, Embryology and Cell Biology* is to put across the morphological aspect of basic medical research and maintain a close reciprocal relationship with the related disciplines by the publication of interesting reviews. These will never be restricted to a one-sided viewpoint, but will be far-reaching and aimed at presentation of the full extent of the morphological bases of medicine.

The appearance of Vol. 100 of *Advances in Anatomy, Embryology and Cell Biology* provides an opportunity for us to thank all the authors who have made their results available for publication in this monograph series. At the same time, thanks are also due to the publishing house and the people there who have kept the series going through what have sometimes been hard times and have always taken good care of it. Thanks are due no less to the present editors and their predecessors, who have never spared themselves in their efforts to make this monograph series interesting and scientifically attractive. We ourselves wish *Advances in Anatomy, Embryology and Cell Biology* continued success over another century.

Würzburg, September 1986

F. Beck, Leicester
W. Hild, Galveston
W. Kriz, Heidelberg
R. Ortmann, Köln
J.E. Pauly, Little Rock
T.H. Schiebler, Würzburg

Acknowledgements

This work was supported by grants from the National Institutes of Health and the National Science Foundation. We wish to thank P. Cleary, J. Henderson, N. Kauper, K. Shuster and G. Whitham for their excellent technical assistance.

Contents

Abbreviations

ab	anterobasal neuroepithelium
AB	anterobasal nucleus
AC	anterior commissure
ACN	nucleus of anterior commissure
AD	anterodorsal hypothalamic nucleus
ah	anterior hypothalamic neuroepithelium
AH	anterior hypothalamus
ahc	anterior hypothalamic neuroepithelium, caudal
ahd	anterior hypothalamic neuroepithelium, dorsal
ahr	anterior hypothalamic neuroepithelium, rostral
ap	anterior periventricular neuroepithelium
AP	anterior periventricular area
AQ	aqueduct
AQi	aqueduct inferior
AQs	aqueduct superior
ar	arcuate neuroepithelium
AR	arcuate nucleus
AV	anteroventral hypothalamic nucleus
AVl	anteroventral hypothalamic nucleus, lateral
bp	base plate domain
bst	bed nucleus of stria terminalis neuroepithelium
BST	bed nucleus of stria terminalis
bt	basal telencephalic neuroepithelium
BT	basal telencephalon
bta	basal telencephalon, anterior hillock neuroepithelium
btp	basal telencephalon, posterior hillock neuroepithelium
bts	subependymal layer of posterior basal telencephalon
c	caudal
cc	cortical (cerebral) neuroepithelium
ce	cerebellar neuroepithelium
CE	cerebellum
cgm	chiasmatic glial matrix
CIL	ciliated ependyma
CON	convoluted ependyma
cp	choroid plexus rudiment
CP	cerebral peduncle
d	dorsal
dB	neuroepithelium of the diagonal band of Broca
dc	neuroepithelium of the dorsocaudal nucleus

DC	dorsocaudal nucleus (posterior area of hypothalamus)
DI	diamond-shaped ventricle
dm	neuroepithelium of the dorsomedial nucleus
DM	dorsomedial nucleus
DMb	dorsomedial nucleus, basal subnucleus
DMd	dorsomedial nucleus, dorsal part
DMv	dorsomedial nucleus, ventral part
dz	differentiating zone
e	eye rudiment
ep	entopeduncular nucleus neuroepithelium
EP	entopeduncular nucleus
et	epithalamic (habenular) neuroepithelium
etr	epithalamic recess
fc	falx cerebri
fdi	diencephalic flexure, inner
fdo	diencephalic flexure, outer
FM	foramen of Monro
FMh	foramen of Monro, horizontal branch
FMv	foramen of Monro, vertical branch
FX	fornix
gp	presumptive neuroepithelium of the globus pallidus (external)
GP	globus pallidus (external)
hbr	habenular recess
hi	hippocampal neuroepithelium
HI	hippocampal formation
hl	lateral habenular neuroepithelium
HL	lateral habenular nucleus
hmc	magnocellular hypothalamic neuroepithelium
HP	habenulopeduncular tract
ibf	interbasal telencephalic fissure
ic	inferior collicular neuroepithelium
IMC	internuclear magnocellular neurons
INT	internal capsule
ir	infundibular recess
is	isthmal neuroepithelium
IS	isthmus
l	lateral
lam	laminated germinal matrix
LAM	laminated ependyma
lh	lateral hypothalamic neuroepithelium
LH	lateral hypothalamus
lt	lamina terminalis
ltr	recess of the lamina terminalis
m	medial
ma	mammillary neuroepithelium
MC	medial mammillary nucleus, central subnucleus
MCO	mammillary commissure
ME	median eminence

MFB	medial forebrain bundle
MG	medial geniculate nucleus
mh	magnocellular hypothalamic neuroepithelium
mi	migratory cells
MI	medial mammillary nucleus, pars intermedialis
MId	medial mammillary nucleus, pars intermedialis dorsalis
MIv	medial mammillary nucleus, pars intermedialis ventralis
ML	lateral mammillary nucleus
MM	medial mammillary nucleus
MMd	medial mammillary nucleus, diffuse region
mr	mammillary recess
MTE	mammillotegmental tract
MTH	mammillothalamic tract
ne	neuroepithelium
OC	optic chiasma
of	optic fissure
ON	optic nerve
or	optic recess
OT	optic tract
pa	preoptic area neuroepithelium
PA	preoptic area
PAL	anterolateral preoptic region
PAM	anteromedial preoptic region
PC	posterior commissure
PE	periventricular field
PF	hypothalamic perifornical nucleus
phd	posterior hypothalamic neuroepithelium, dorsal
PHd	posterior hypothalamus, dorsal
phv	posterior hypothalamic neuroepithelium, ventral
PHv	posterior hypothalamus, ventral
PI	pituitary gland (or Rathke's pouch)
PIM	intermediate preoptic region
pm	premammillary neuroepithelium
PM	premammillary nuclei
pmd	neuroepithelium of the dorsal premammillary nucleus
PMD	dorsal premammillary nucleus
pmv	neuroepithelium of the ventral premammillary nucleus
PMV	ventral premammillary nucleus
pn	posterior nuclei neuroepithelium
po	pontine neuroepithelium
PO	pons
pp	posterior periventricular neuroepithelium
PP	posterior periventricular area
PPD	posterodorsal preoptic region
PPL	posterolateral preoptic region
PPV	posteroventral preoptic region
pr	preoptic recess
prp	preoptic periventricular neuroepithelium
PS	superior (median) preoptic region

pt	pretectal neuroepithelium
PT	pretectum
PV	hypothalamic paraventricular nucleus
PVa	anterolateral subnucleus of PV
PVc	posterior subnucleus of PV
PVm	medial subnucleus of PV
PVp	parvocellular subnucleus of PV
pz	proliferative zone
r	rostral
re	neuroepithelium of the retina
rt	reticular protuberance
sc	superior collicular neuroepithelium
SC	superior colliculus
SD	supraoptic decussation
se	septal neuroepithelium
SE	septum
SF	septofimbrial nucleus
SFO	hypothalamic suprafornical nucleus
SM	supramammillary nucleus
SMl	supramammillary nucleus, pars lateralis
SMm	supramammillary nucleus, pars medialis
so	supraoptic neuroepithelium
SO	supraoptic nucleus
SOd	supraoptic nucleus, dorsolateral region
SOv	supraoptic nucleus, ventromedial region
sp	surface plate domain
st	subthalamic neuroepithelium
ST	subthalamic nucleus
STa	accessory subthalamic nucleus
STc	subthalamic nucleus, caudal part
STi	subthalamic nucleus, intermediate part
STM	stria medullaris
STr	subthalamic nucleus, rostral part
STT	stria terminalis
su	suprachiasmatic nucleus neuroepithelium
SU	suprachiasmatic nucleus
SUb	suprachiasmatic nucleus, basal subnucleus
sud	neuroepithelium of the dorsomedial subnucleus of SU
SUd	suprachiasmatic nucleus, dorsomedial subnucleus
suv	neuroepithelium of the ventrolateral subnucleus of SU
SUv	suprachiasmatic nucleus, ventrolateral subnucleus
sz	synthetic zone
ta	anterior thalamic neuroepithelium
TA	anterior thalamus
tan	tanycytic germinal matrix
TAN	tanycytic ependyma
td	dorsal thalamic neuroepithelium
TD	dorsal thalamus
tdd	telencephalic-diencephalic fissure, dorsal

tds	dorsal thalamic subependymal layer
tdv	telencephalic-diencephalic fissure, ventral
te	tegmental neuroepithelium
TE	tegmentum
tel	lower tegmental neuroepithelium
TEl	lower tegmentum
tem	medial tegmental neuroepithelium
teu	upper tegmental neuroepithelium
TEu	upper tegmentum
th	thalamic neuroepithelium
TH	thalamus
TL	telencephalon
tp	posterior thalamic neuroepithelium
tr	reticular thalamic neuroepithelium
TR	reticular thalamus
TS	nucleus triangularis septi
TU	tuberal magnocellular nucleus
v	ventral
vl	lateral ventricle
vlm	medial horn of the lateral ventricle
vm	neuroepithelium of the ventromedial nucleus
VM	ventromedial nucleus of the hypothalamus
VMb	ventromedial nucleus, basal subnucleus
VMd	ventromedial nucleus, pars dorsalis
VMv	ventromedial nucleus, pars ventralis
v3	third ventricle
v3d	third ventricle, dorsal (thalamic) domain
v3v	third ventricle, ventral (hypothalamic) domain
v4	fourth ventricle
zi	zona incerta neuroepithelium
ZI	zona incerta
ZIf	zona incerta, far-lateral part
ZIl	zona incerta, lateral part
ZIm	zona incerta, medial part

1 Introduction

The hypothalamus is a vitally important component of the vertebrate nervous system, being involved in the regulation of the pituitary gland, the autonomic nervous system, and the central and peripheral motor systems; and implicated in such functions as reproduction, feeding, drinking, aggression, vigilance, and some other not clearly understood motivational processes. The development of the hypothalamus is generally recognized as an important field of inquiry, and many studies are available that deal with the biochemical, electrophysiological, morphological, and behavioral maturation of one or another component of this unique brain region. However, comprehensive studies of the development of the hypothalamus are scant. Some of the early attempts have been reviewed by Diepen (1962). Among the more recent studies, some in the context of descriptions of the development of the entire diencephalon, are those by Rose (1942) in the rabbit; Luyendijk (1944) and Niimi et al. (1962) in the mouse; Auer (1951) and Keyser (1972) in the hamster; Ströer (1956), Coggeshall (1964), and Hyyppä (1969) in the rat; and Gilbert (1953), Kahle (1956), and Richter (1965) in man. Thymidine radiography was utilized in the study by Ifft (1972) in the rat, Shimada and Nakamura (1973) in the mouse, and Wyss and Sripanidkulchai (1985) in the cat.

In a recent study (Altman and Bayer 1978b) we have made an attempt to correlate the development of the rat hypothalamus, using paraffin-embedded normal and X-irradiated embryos, with data obtained of the time origin of hypothalamic neurons with long-survival (adult) thymidine radiography (Altman and Bayer 1978a). In the present reexamination of the embryonic development of the rat hypothalamus we are using a new collection of paraffin-embedded and methacrylate-embedded short-survival and sequential (embryonic) thymidine radiograms to better delineate the site of origin, path of migration, and settling pattern of neurons of the different structural components of the hypothalamus. This ontogenetic approach, combined with available data from long-survival thymidine radiograms, has allowed us to propose a rational subdivision of the hypothalamus, identify several new hypothalamic nuclei, and trace the germinal source of several hypothalamic structures. It will be shown that the neurons of the hypothalamus are released from the third ventricle neuroepithelium in three successive waves, giving rise to three structurally and functionally discrete longitudinal tiers. These will be called the reticular hypothalamus, the core hypothalamus, and the midline hypothalamus. Conspicuous early landmarks along the third ventricle neuroepithelium permit the subdivision of the core hypothalamus into four transverse segments, the preoptic, the anterior and posterior hypothalamic, and the mammillary. Different autoradiographic labeling patterns within relatively homogeneous structures, and the circumstance that nuclear boundaries are clearer in many regions in the fetus than in adults, have made it possible to identify hitherto unnamed hypothalamic nuclei. Finally, by identifying the migratory paths of neurons of several hypothalamic structures, we have been able to delineate with some confidence the neuroepithelial zones, or germinal mosaics, where they originate.

2 Materials and Methods

Four collections of thymidine radiograms were used. The first will be referred to as *short-survival radiograms*. In this series the embryos or fetuses were killed 2 h after their mothers were injected with [^{3}H]thymidine in order to study the sites of cell proliferation, and their magnitude, as a function of embryonic age at injection. The second collection constitutes the *sequential radiograms*. This series consists of groups that were killed at sequential daily intervals after injection on a particular day in order to trace cell migration. The third series is referred to as *fetal long-survival radiograms*. In this series the fetuses were killed 1–2 days before birth to study the settling patterns and nuclear organization of neurons. Finally, use was made of an older series of *adult long-survival radiograms* in which double injections of [^{3}H]thymidine were made at selected ages to allow the quantitative specification of the time of origin of hypothalamic neurons (Altman and Bayer 1978a).

All the specimens were obtained from laboratory-bred Purdue-Wistar rats, using the following standardized procedures. Female rats were mated in the evening and vaginal smears were done the next morning. Those found sperm positive were considered to have reached gestational (embryonic) day 1 (E1) and were isolated in maternity cages. On selected gestational days the dams were injected with [^{3}H]thymidine subcutaneously between 9:00 and 11:00 a.m., 5 μc/g body weight, specific activity of the radiochemical 6.0 Ci/m*M*. The dams were anesthetized before the embryos and fetuses were removed, and the specimens were blocked immediately coronally, sagittally, or horizontally, and immersed in Bouin's fluid. After 24 h, the material was transferred for storage into 10% neutral formalin. Subsequently the blocks were embedded either in paraffin or methacrylate. The paraffin blocks were serially sectioned at 6 μm and the methacrylate blocks at 3 μm, and every tenth section was saved. Successive sections were either stained with cresyl violet and hematoxylin-eosin for examination without nuclear emulsion or were prepared for autoradiography. The latter were coated with Kodak NTB-3 emulsion in the dark, exposed for 90 days with a desiccant, developed in Kodak D-19, and stained with hematoxylin-eosin.

Short-Survival Radiograms. This series consists of 94 paraffin- or methacrylate-embedded embryos or fetuses from dams that were injected with a single dose of [^{3}H]thymidine at daily intervals between days E12 and E21, and were removed and fixed 2 h after injection. The number of specimens examined in the different age groups ranged from 6 to 12.

Sequential Radiograms. This series consists of 254 paraffin- and methacrylate-embedded embryos and fetuses whose mothers were injected with a single dose of [^{3}H]thymidine at daily intervals between days E12 and E21. Within each injection group, the embryos and fetuses were removed and fixed at daily intervals after the injection either up to day E22 or, in the case of late fetal injections, up to postnatal day 5 (P5). A minimum number of six specimens were examined in every relevant survival group.

Fetal Long-Survival Radiograms. This series consists of 231 paraffin- and methacrylate-embedded fetuses or young pups whose mothers received [^{3}H]thymidine at daily intervals from days E12 to E21, and were killed either perinatally (days E20–E22) or postnatally (P5). In the present study only the perinatally killed specimens were utilized.

Adult Long-Survival Radiograms. This older collection was previously used in several of our studies to date the time or origin of neurons in the central and peripheral nervous systems of adult rats. Groups of pregnant rats were injected with two successive daily doses of [^{3}H]thymidine from gestational day E13 (E13 + 14) until the day before birth (E21 + 22) with a single day delay between the groups (E13 + 14, E14 + 15, E15 + 16 ...). All the animals were killed at 60 days of age (P60). This is the progressively delayed comprehensive labeling procedure which allows the quantification of the proportion of neurons generated on a single day in a selected brain structure. The dating of the time of origin of neurons of the hypothalamus on the basis of this material has been previously published (Altman and Bayer 1978a).

3 An Overview of the Organization of the Hypothalamus

The hypothalamus was not identified as a distinct brain structure until the end of the nineteenth century. Meynert (1872) considered it to be a part of the thalamus, using the term *centrales Höhlengrau.* Forel (1877) described a *regio subthalamica*, which included both the ventral thalamus and the hypothalamus proper. The term "hypothalamus" (which means the same as subthalamus) was introduced by His (1893). The subsequent attempts to delineate and subdivide the hypothalamus seem to have been guided by two different approaches, one based on embryonic considerations, and the other on structural and functional criteria in adults. The ontogenetic approach was begun by His (1893), who subdivided the diencephalon into three parts, the epithalamus, the thalamus, and the hypothalamus. The other tradition may have been started by Herrick (1910), who, in an investigation of the diencephalon of mature amphibians (urodeles and anurans), identified three fissures of the third ventricle, the sulcus diencephalicus dorsalis, the sulcus diencephalicus medius, and the sulcus diencephalicus ventralis (Fig. 1 A). Herrick used these three landmarks, as seen in coronal sections, to distinguish four dorsoventrally arranged divisions of the diencephalon, the epithalamus, the dorsal thalamus, the ventral thalamus, and the hypothalamus. Thus the hypothalamus of His became subdivided into two parts with the ventral thalamus (or the subthalamus of other authors) no longer considered as part of the hypothalamus. Subsequent work by Kuhlenbeck and his associates (summarized in Kuhlenbeck 1973 pp. 397–401) confirmed the presence of the three diencephalic sulci in a variety of mature anamniotes (cyclostomes, fishes, and amphibians), but in amniotes (reptiles, birds, and mammals) these sulci could be seen only at some transient stage of embryonic development (Fig. 1 B, C). Somewhat inconsistently, perhaps guided by ontogenetic considerations, Kuhlenbeck (1977 pp. 301–302) seems to have reverted to the classification of His when he included the subthalamic nucleus of Luys (a major component of the ventral thalamus) with the "dorsal (hypothalamic) and entopeduncular complex." Rose (1942) in his study of the rabbit diencephalon seems, likewise, to have continued the tradition begun by His of the threefold division of the diencephalon because he included the subthalamic nucleus with what he called the "dorsal hypothalamic area." But this inclusive classification has not been adopted by most anatomists, possibly because such structures as the subthalamus, the zona incerta, and the entopeduncular nucleus are implicated in a class of functions (motor or "extrapyramidal") that is different from the functions of "typical" hypothalamic structures (visceral and hormonal).

Any attempt usefully to classify the hypothalamus has to deal with three related issues: (a) the identification and naming of its discrete components, (b) the ordering of these structures into larger, preferably functionally unified

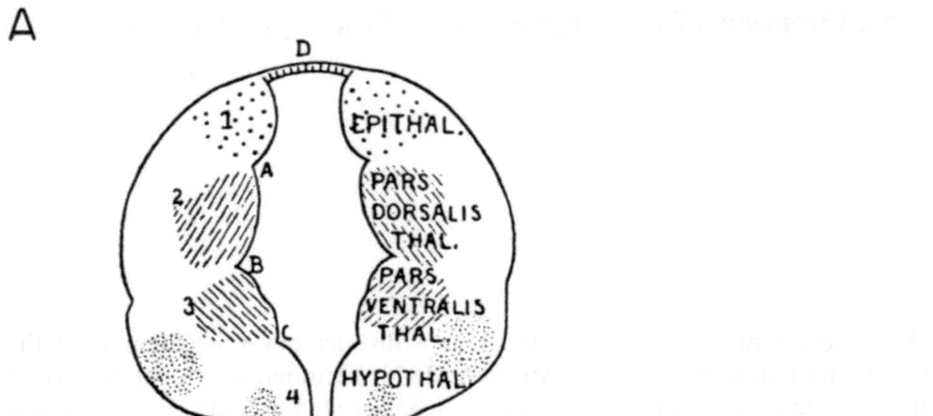

Fig. 1. A The proposed division of the diencephalon by Herrick (1910), based on mature amphibians. *A*, sulcus diencephalicus dorsalis; *B*, sulcus diencephalicus medius; *C*, sulcus diencephalicus ventralis. **B**, **C** Proposed division of the diencephalon by Kuhlenbeck (1973 p. 461), based on studies of the development of human embryos. *cl*, subthalamic nucleus; *et*, epithalamus; *hy*, hypothalamus; *ld*, dorsal nucleus of the lateral geniculate body; *lv*, ventral nucleus of the lateral geniculate body; *ma*, mammillary body; *sd*, sulcus diencephalicus dorsalis; *sm*, sulcus diencephalicus medius; *sv*, sulcus diencephalicus ventralis; *td*, dorsal thalamus; *tv*, ventral thalamus

separate groups or regions, and (c) the delineation of the entire system by determining what structures and regions are or are not parts of the "hypothalamus proper." A common criterion for identifying a discrete brain structure is the delineation of a uniform cell grouping; examples in the hypothalamus are the rather well demarcated supraoptic, paraventricular, or suprachiasmatic nuclei. But these are exceptions. In most regions the identified nuclei, such as the dorsomedial and ventromedial nuclei, have neither a uniform cell composition nor clear boundaries. In still other regions cell groupings are not identifiable at all and anatomists refer to vaguer structures, such as the preoptic area or the lateral hypothalamic area. The simplest criterion for the grouping of discrete structures into larger units is some obvious topographic landmark or feature. Examples are the preoptic area, designating the region lying ahead of the optic chiasma, or the mammillary nuclei, comprising structures enclosed within the mammillary body. Another criterion may be directional, such as medial versus lateral hypothalamus, or anterior versus posterior hypothalamus. Because the functional utility of such classifications are not obvious, physiological criteria are sometimes employed. Examples are the physiotrophic hypothalamus or the sexually dimorphic hypothalamus. The criteria to be employed to demarcate the boundaries of the hypothalamus proper are rarely made explicit. Most anatomists include the preoptic area as part of the hypothalamus, but others (for instance, Le Gros Clark 1938) have excluded it. The preoptic area certainly blends into the telencephalon rostrally and criteria are needed to draw its boundaries if it is to be considered a component of the diencephalic hypothalamus. Most anatomists do not consider the subthalamic nucleus or the entopeduncular nucleus as part of the hypothalamus but some (for instance, Kuhlenbeck 1977) have argued for their inclusion on the basis of ontogenetic and phylogenetic considerations. Certainly there is need for an explicit criterion to delineate the dorsal boundary of the hypothalamus along the third ventricle.

An early attempt to subdivide the hypothalamus was made by Malone (1910). In addition to the supraoptic nucleus (then called the ganglion opticum basale) and the paraventricular nucleus (which many early anatomists referred to as the filiform nucleus), Malone distinguished several components of the mammillary body and identified the tuberal region and the substantia grisea of the third ventricle. The subdivisions currently in use began with the work of Gurdjian (1927) and Krieg (1932) in the rat. A bibliographic summary of subsequent studies in a great many species, together with a review of various regional and nuclear subdivisions, is presented by Diepen (1962).

The simplest of recent regional subdivisions of the hypothalamus is that of Christ (1969) into three "groups." This division is along the longitudinal axis into an anterior, middle, and posterior group. The anterior group includes several nuclei of the preoptic area, and the suprachiasmatic, anterior hypothalamic, paraventricular, and supraoptic nuclei. The middle group includes the lateral hypothalamic and dorsal hypothalamic areas, and the tuberomammillary, lateral tuberal, arcuate, ventromedial, and dorsomedial nuclei. The posterior group includes several mammillary nuclei, and the perifornical nucleus, the nucleus intercalatus, and the posterior periventricular area. This directional classification uses neither the dorsoventral axis nor the lateromedial axis for further subdivision of components of the hypothalamus. Prior to this attempt, Rioch et al. (1940) presented a somewhat more complex classification into six "regions," one that does justice to the obvious transverse division of the hypothalamus into a medial nuclear component and a lateral reticular component. Rioch et al. subdivided the anterior hypothalamus into a preoptic region and a supraoptic region, distinguished the middle (tuberal) and posterior (mammillary) regions, and added the periventricular and the lateral subdivisions of the hypothalamus. The most complex and detailed recent classification, with extensive illustrations, is that of Bleier et al. (1979). Bleier and her coworkers identify 40 nuclei and areas of the rat hypothalamus and order them into a system of regions, divisions, and zones. The "regions" are along the anteroposterior and dorsoventral axes, and are called the anterior (including the preoptic), tuberal, dorsal, and mammillary regions. The two "divisions" are in the horizontal plane, called the medial and the lateral; and the medial division is further subdivided into a medial and periventricular "zone." Bleier et al. list several structures that are said to derive from the hypothalamic anlage but are not considered to be part of it; these are listed as the bed nucleus of the stria terminalis, the nucleus of the diagonal band of Broca, globus pallidus, entopeduncular nucleus, zona incerta, nucleus of the fields of Forel, and the subthalamic nucleus of Luys.

In the present approach ontogenetic considerations are the primary guides both for the delineation of the boundaries of the hypothalamus and its parcellation. But we can reconcile this approach with a directional subdivision of the hypothalamus that makes use of landmarks along the dorsoventral, lateromedial, and anteroposterior planes. Because of developmental considerations to be described in detail later, we propose to follow the simpler subdivision of the diencephalon by His and include the ventral thalamus (or subthalamus) with the hypothalamus. All structures lying beneath the thalamus are considered to belong to the hypothalamus. It will suffice here to mention the two major ontogenetic considerations. First, it will be shown that the subthalamic nucleus of Luys, a major component of the classical ventral thalamus, originates from the region of the mammillary recess neuroepithelium, that is, from a germinal site that is considered by all to be part of the anlage of the hypothalamus. Second, evidence will be presented that the neurons of the lateral hypothalamic area do not originate from the classical hypothalamic anlage but from a dorsal region that also produces the neurons of the entopeduncular nucleus and the zona incerta. Thus, developmentally the lateral hypothalamic area is closely affiliated with structures considered to constitute the bulk of the ventral thalamus.

Figure 2A schematically presents the proposed boundaries and primary divisions of the hypothalamus at a midcoronal level. The three transverse divisions are the reticular hypothalamus, the core hypothalamus, and the midline hypothalamus. The reticular hypothalamus is situated directly beneath the thalamus and spreads from dorsomedial to ventrolateral in the coronal plane and from the preoptic area to the mammillary body in the anteroposterior plane. Its major components are the zona incerta, the entopeduncular nucleus, and the lateral hypothalamic area. The neurons of all these structures are generated early during embryonic life and constitute what we consider the "first wave" of hypothalamic neurons. The "second wave" constitutes the hypothalamic core. The hypothalamic core is divisible into four longitudinal or anteroposterior components (Fig. 2B): the preoptic area, the anterior hypothalamus, the posterior hypothalamus, and the mammillary body. Embryonically each can be related to a different segment of the hypothalamic neuroepithelium. Major structural components of the anterior hypothalamus are, rostrally, the anterobasal nucleus and the paraventricular nucleus, and caudally, the dorsomedial and ventromedial nuclei; and of the posterior hypothalamus, the premammillary and supramammillary nuclei, the tuberal magnocellular nucleus, and the subthalamic nucleus. The "third wave" of hypothalamic neurons form the midline hypothalamus. These late-produced neurons remain in the vicinity of the third ventricle. Specialized components of the midline hypothalamus are the suprachiasmatic

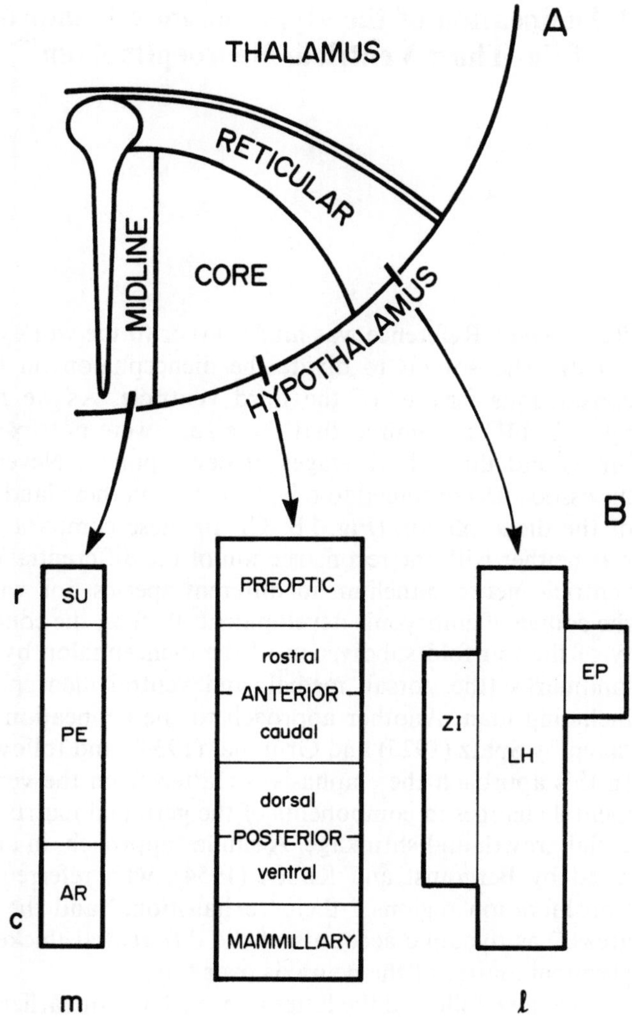

Fig. 2. A Schematic illustration of the three primary divisions (tiers) of the hypothalamus in the coronal plane. **B** Major components of the three divisions of the hypothalamus in the anteroposterior plane

nucleus anteriorly and the arcuate nucleus posteriorly. The extended periventricular area makes up its bulk. In addition to the possibility of providing a rational subdivision of the hypothalamus, the developmental approach offers additional advantages. Hypothalamic structures that cannot be identified in the mature diencephalon because the boundaries of the cell groups are vague, or are altogether obliterated due to cell dispersion and the penetration of various fiber bundles, may be easier to delineate in embryos with a tighter cell packing density and paucity of fibrous elements. Thymidine radiographic labeling offers an additional aid because many cell groups are distinguishable from one another by their different labeling patterns (different times of origin and/or different neurogenetic gradients).

4 Delineation of the Hypothalamic Primordium of the Third Ventricle Neuroepithelium

Background. Reference was made earlier to the work of His (1893) and Herrick (1910), who sought to divide the diencephalon on the basis of some of the conspicuous fissures of the third ventricle. As we noted, Kuhlenbeck (1973 pp. 397–401) recognized that these sulci were not fixed landmarks in all vertebrates and through all stages of development. Nevertheless, Kuhlenbeck and his associates continued to rely primarily on these landmarks for the parcellation of the diencephalon (Fig. 1B, C). In these comparative attempts, the concern was neither with the reconstruction of the differential configuration of the third ventricle neuroepithelium in different species nor with its transformations in the course of embryonic development. Rather, the concern was to test the validity of the fourfold subdivision of the diencephalon by the presence of the three landmarks (the dorsal, medial, and ventral diencephalic sulci) in all species, including man. Another approach to the delineation of the diencephalon was taken by Spatz (1925) and Grünthal (1930), and followed later by Kahle (1956). In this approach the emphasis is shifted from the ventricular sulci to developmental changes in components of the germinal matrix, in particular their differential growth and shrinkage. A similar approach, in another context, was advocated by Bergquist and Källén (1954), who referred to the intensification of "proliferation regions," their "exhaustion," and the formation of "migration areas," as dynamic accounts of the differential thickening and thinning of the germinal matrix of the neuraxis over time.

We have followed the latter approach in our earlier study of the development of the diencephalon (Altman and Bayer 1978b, 1979b). Although we could identify the dorsal, medial, and ventral diencephalic sulci in midcoronal sections of the third ventricle in day E13 rats, we have pointed out that the sulcal pattern is different at this age rostrally and caudally, and that it changes drastically throughout the third ventricle on the subsequent days. The sulci, we have argued, are the shifting boundaries of the neuroepithelial "lobes" and "lobules" that produce neurons for the major divisions of the diencephalon. We have identified three lobes, the dorsal, medial, and ventral, and proposed that they are sources of neurons of the thalamus, ventral thalamus, and hypothalamus, respectively. The inferior lobule was postulated to be the site of origin of neurons of the hypophysiotropic area of the hypothalamus (arcuate nucleus and other structures) and the mammillary recess neuroepithelium of the mammillary body.

In the present study we continue with this approach but modify and extend the parcellation of the third ventricle neuroepithelium on the basis of an examination of a large series of short survival and sequential thymidine radiograms that have now become available to us. We begin with a simplification. We postulate that there are only two major divisions of the third ventricle neuroepithelium: one being the source of neurons of the thalamus, the other of the hypothalamus. These two neuroepithelial divisions can be distinguished from each other throughout embryonic development and along the entire length of the third ventricle (where both are present). The next step greatly complicates matters. It begins with the hypothesis that the diencephalic neuroepithelium is a mosaic of heterogeneous neuroepithelial zones, each zone being a proliferative center composed of a cell line destined to produce neurons for a specific diencephalic structure. But the attempt to delineate these neuroepithelial zones is made difficult by two circumstances. First, the neuroepithelium is not truly a discontinuous mosaic but a continuous sheet of cells which, during the early stages of embryonic development, extends from one end of the neuraxis to the other. The mosaic composition of the neuroepithelium becomes manifest only if studied serially over time because the neurons destined for different brain structures multiply at different times and rates and produce more active and less active foci along this continuum. Since cell mitosis in the neuroepithelium occurs near the lumen (Sauer 1935; Sidman et al. 1959), a high level of proliferative activity requires the lengthening of the ventricular surface. This can be accomplished by either an evagination or invagination of the neuroepithelium. The neuroepithelial zones are thus delineated in the undulating neuroepithelium as regional concavities and convexities. Because these inward and outward bulgings vary in their configuration in three-dimensional space, their delineation requires serial reconstruction. The second difficulty in this attempt of mapping the third ventricle neuroepithelium is the uncertainty as to what represents a discrete diencephalic structure that ought to have a discrete neuroepithelial zone. Therefore, the examination of the changing configuration of the diencephalic neuroepithelium has to be coupled with a study of cell migration from the neuroepithelial sites and patterns of cell settling, and with a structural and functional consideration of the divisions and subdivisions of the diencephalon in adults.

The parcellation of the thalamic division of the third ventricle neuroepithelium will be attempted in the next monograph of this series dealing with the embryonic development of the thalamus (Altman and Bayer in preparation). In the present delineation and parcellation of the hypothalamic neuroepithelium we shall make use of identifiable flexures, deep recesses, as well as the regional evaginations and invaginations, and thickenings and thinnings of the neuroepithelium over time in relation to the chronology of the time of origin of neurons of different hypothalamic structures.

Results. The configuration of the lumen of the third ventricle and its neuroepithelial lining in relation to the rest of the embryonic brain is illustrated in four sagittally sectioned radiograms from medial to lateral in a rat that received [^{3}H]thymidine at about 9:00 a.m. on *day E13* and was killed 2 h later (Figs. 3, 4). A line drawn from the depth of the dorsal telencephalic-diencephalic fissure (tdd) to the ventral telencephalic-diencephalic fissure (tdv) at the posterior boundary of the basal telencephalic neuroepithelium (bt) demarcates the anterior boundary of the third ventricle (v3). Another line drawn from the boundary between the pretectal and dorsal thalamic neuroepithelium to the inner diencephalic flexure (fdi) ventrally delineates the posterior boundary of the third ventricle. The broken line drawn from the fourth ventricle (v4) to the lateral ventricle

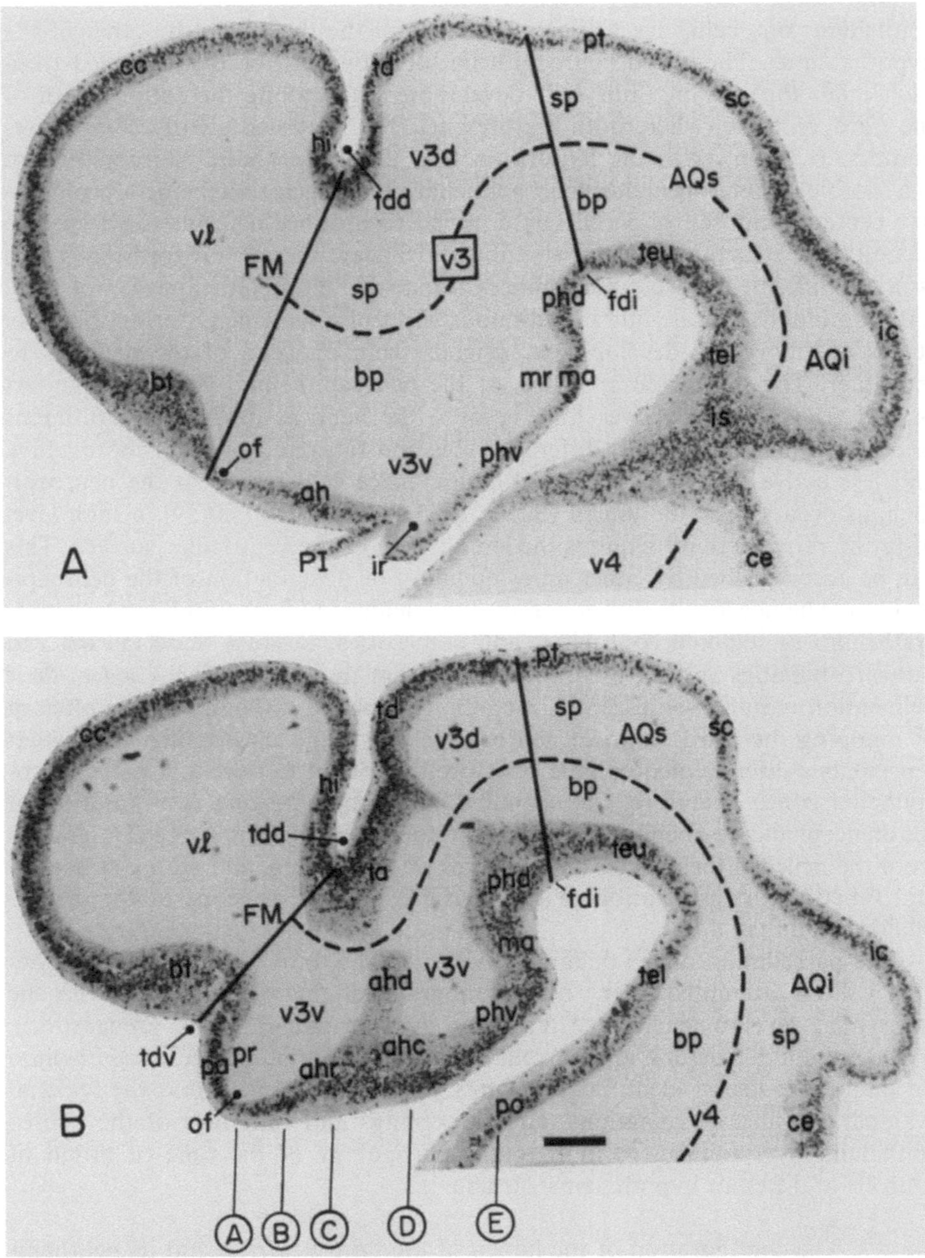

Figs. 3 and 4. Sagittal thymidine radiograms of the brain from medial (Fig. 3A) to lateral (Fig. 4B) from a rat labeled on the morning of day E13 and killed 2 h later. *Solid lines* delineate the rostral and caudal boundaries of the third ventricle and the diencephalic neuroepithelium. *Broken lines* separate the surface plate and the basal plate of the neuroepithelium and, within the domain of the diencephalon, the thalamic rudiment from the hypothalamic rudiment. The vertical lines *A–E* correspond to the coronal sections *A–E* in Fig. 5. Methacrylate. *Scales*, 200 μm

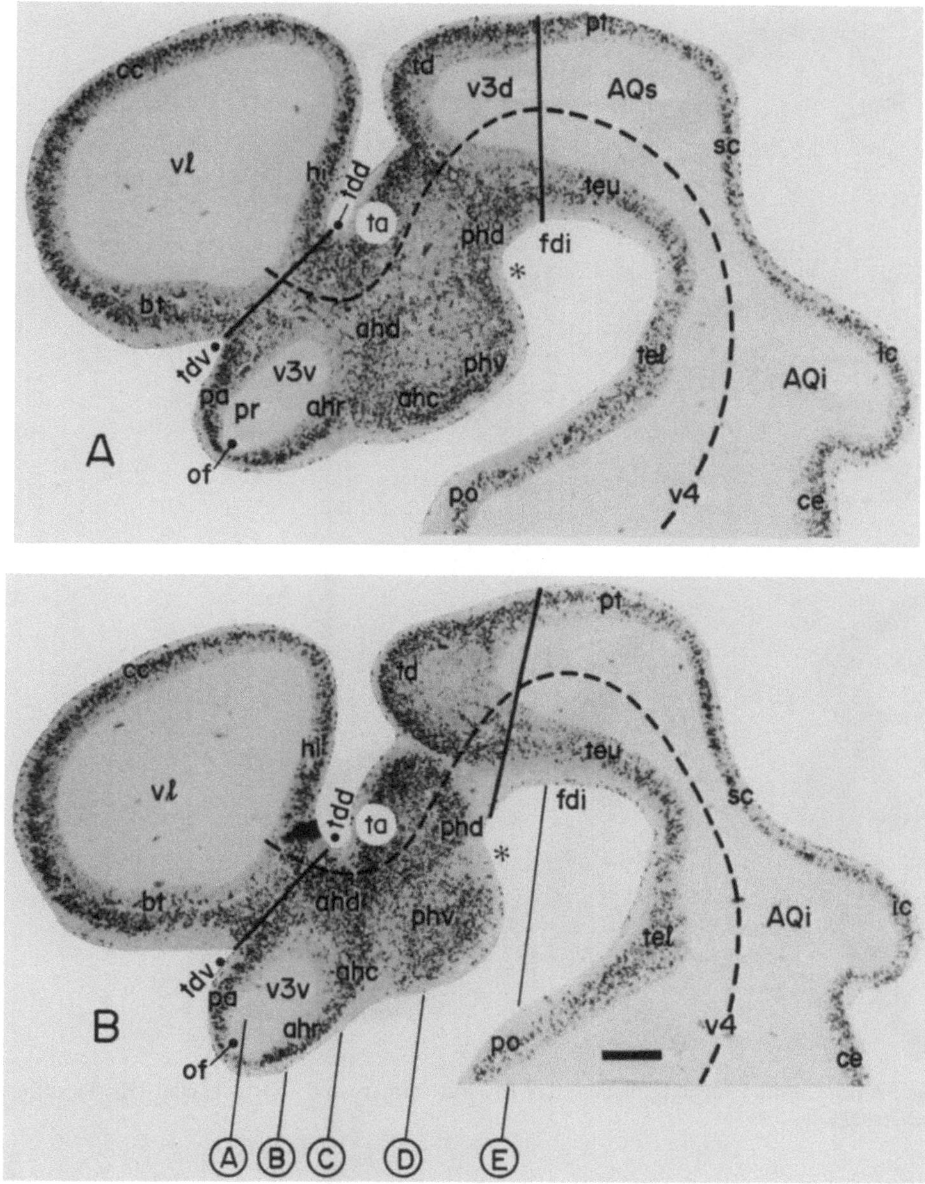

Fig. 4. Legend see Fig. 3

(vl) to separate the surface plate neuroepithelium (sp; which gives rise to afferent structures) from the basal plate neuroepithelium (bp; which is a source of efferent structures) divides the dorsal third ventricle neuroepithelium (v3d) from the ventral third ventricle neuroepithelium (v3v). The former is the primordium of the thalamus, the latter of the hypothalamus.

In the hypothalamic domain of the third ventricle there are three recesses medially (Fig. 3A–B) from caudal to rostral: the mammillary recess (mr), the infundibular recess (ir), and the preoptic recess (pr). Another landmark is the optic fissure (of), which can be traced laterally throughout the anteroventral

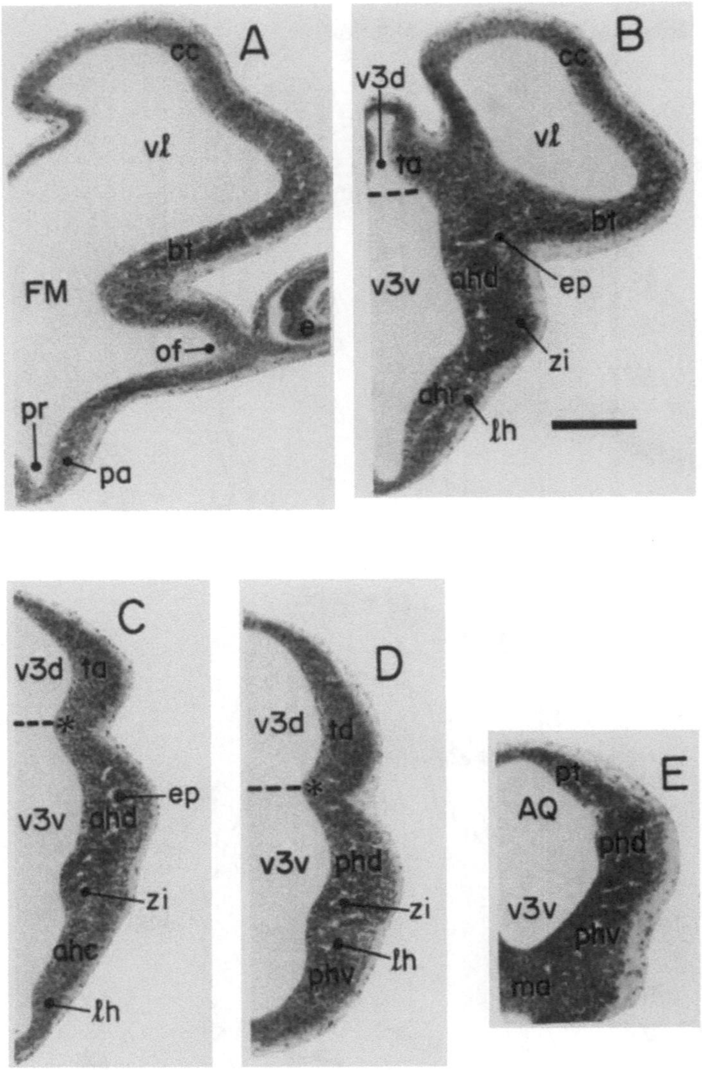

Fig. 5 A–E. Coronal sections from a day-E13 rat from rostral (**A**) to caudal (**E**). Paraffin. *Scale*, 400 μm

hypothalamus. These landmarks are used to delineate components of the hypothalamus along the longitudinal axis. The neuroepithelium situated between the inner diencephalic fissure (fdi) and the mammillary recess is called the *dorsal region of the posterior hypothalamic rudiment* (phd). The neuroepithelium beneath it, lining the mammillary recess, is the *rudiment of the mammillary complex* (ma). The neuroepithelium between the mammillary recess and the infundibular recess is the *ventral region of the posterior hypothalamic rudiment* (phv). Laterally (Fig. 4 A, B), where the mammillary recess is not present, a fissure (asterisk) separates the contiguous dorsal and ventral regions of the posterior hypothalamic rudiment. The hillock between the infundibular recess and the optic fissure is the *rudiment of the anterior hypothalamus* (ah in Fig. 3 A). This hillock grows in height laterally and is divisible into two parts: the half adjacent to the infundi-

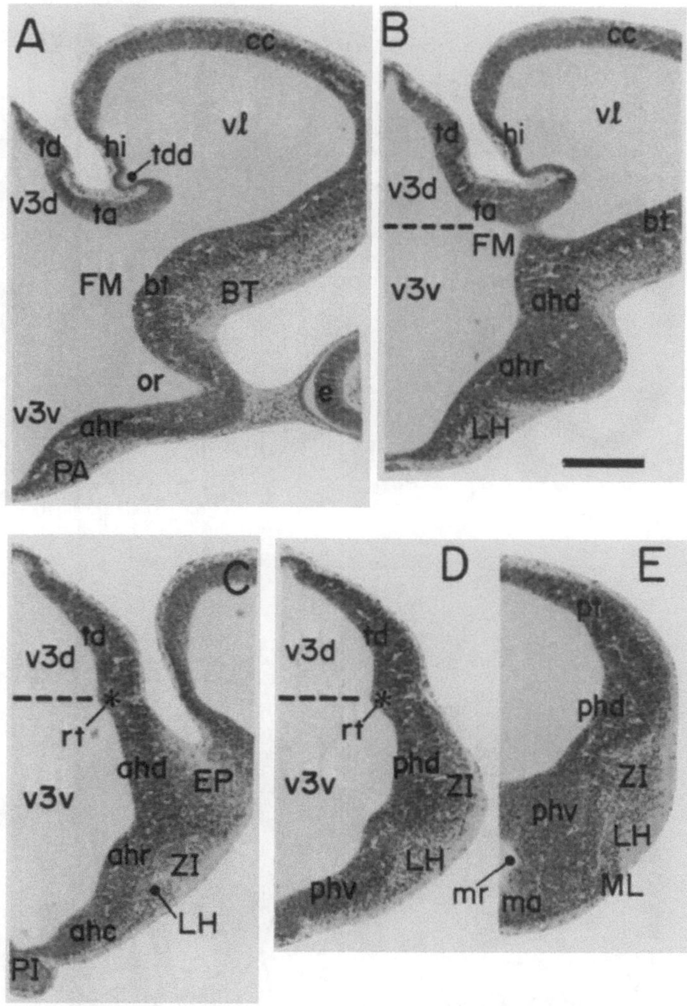

Fig. 6 A–E. Coronal sections from a day-E14 rat from rostral (**A**) to caudal (**E**). Paraffin. *Scale*, 400 μm

bular recess will be called the *caudal region of the anterior hypothalamic rudiment* (ahc), and the half adjacent to the optic fissure is the *rostral region of the anterior hypothalamic rudiment* (ahr). The peak of the anterior hypothalamic rudiment will be referred to as the *dorsal region of the anterior hypothalamic rudiment* (ahd); this region becomes apposed to the thalamus. The *rudiment of the preoptic area* (pa) is situated rostral to the optic fissure and is separated from the basal telencephalon by the ventral telencephalic-diencephalic fissure (tdv).

The configuration of the lumen and of the neuroepithelial lining of the third ventricle of a day-E13 rat in coronal sections is illustrated in Fig. 5. The five levels shown from rostral to caudal (Fig. 5A–E) roughly correspond to the vertical lines A–E in Figs. 3B and 4B. The preoptic area neuroepithelium ventral to the optic fissure and lateral to the preoptic recess is seen most rostrally (Fig. 5A). More posteriorly (Fig. 5B), the neuroepithelium of the anterior thala-

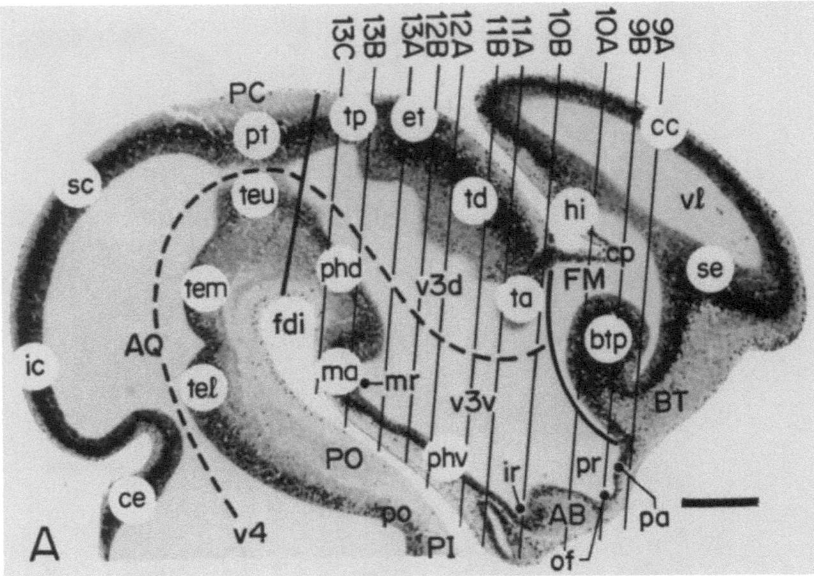

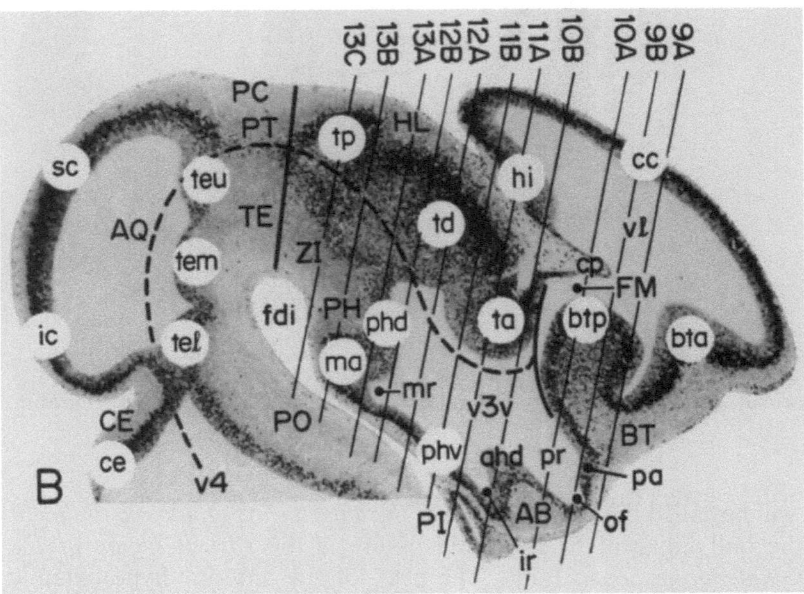

Figs. 7 and 8. Sagittal thymidine radiograms of the brain from medial (Fig. 7A) to lateral (Fig. 8B) from a rat labeled on the morning of day E15 and killed 2 h later. *Solid and broken lines* are used as in Figs. 3, 4. The vertical lines *9A–13C* correspond to the coronal sections in Figs. *9A–13C*. Paraffin. *Scales*, 400 μm

mus caps the third ventricle. Beneath it are the dorsal and rostral regions of the anterior hypothalamic rudiment. At the next two caudal levels (Fig. 5C, D), the reticular protuberance (asterisk) separates the thalamic and hypothalamic rudiments. [The reticular protuberance is the source of neurons of the reticular thalamus (Altman and Bayer in preparation).] The relatively small area occupied

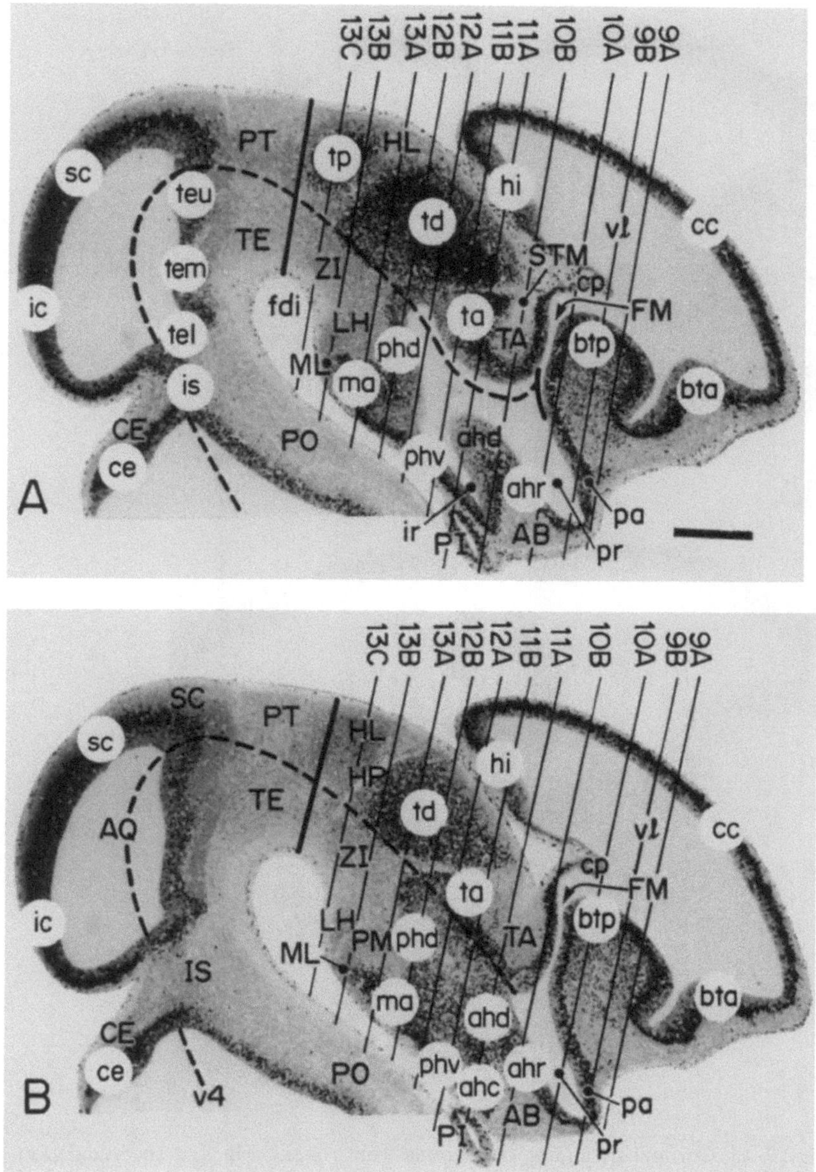

Fig. 8. Legend see Fig. 7

by the thalamic rudiment on day E13 agrees with the evidence that thalamic neurogenesis (with the major exception of the reticular thalamus and the lateral habenular nucleus) does not begin until day E14, and peak production of thalamic neurons is on days E15 and E16 (Altman and Bayer 1979a: Figs. 6, 10, 12). In contrast, the hypothalamic rudiment is extensive.

We shall present evidence later that at this stage of development the first wave of hypothalamic neurons, those of the reticular hypothalamus, are generated. The upper lobule of the dorsal region of the anterior hypothalamic rudi-

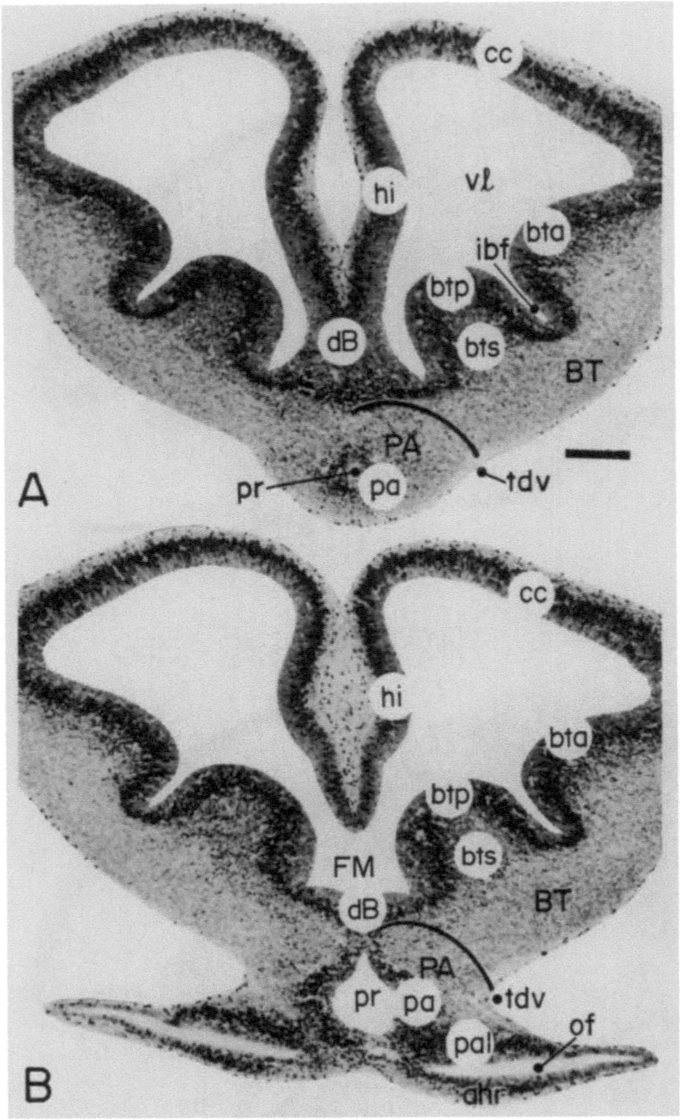

Figs. 9–13. Coronal thymidine radiograms from rostral (Fig. 9A) to caudal (Fig. 13C) from a rat labeled on the morning of day E15 and killed 2 h later. In these and the following illustrations *abbreviations in lowercase letters* refer to germinal zones, *those in capital letters* to differentiating zones. Paraffin. *Scales*, 200 μm

ment is the source of entopeduncular neurons (ep; Sect. 5.2). The zona incerta neurons (zi; Sect. 5.3) are derived from the middle lobule of the dorsal regions of the anterior and posterior hypothalamic rudiments. Finally, the neurons of the lateral hypothalamic area (lh; Sect. 5.1) derive from the ventral regions of the rostral and caudal anterior hypothalamic, and of the posterior hypothalamic, rudiments. By *day E14*, differentiating neurons of the lateral hypothalamic area (LH in Fig. 6B–E), of the zona incerta (ZI in Fig. 6C–E), and of the entopeduncular nucleus (EP in Fig. 6C) form lightly staining cell clusters. Other

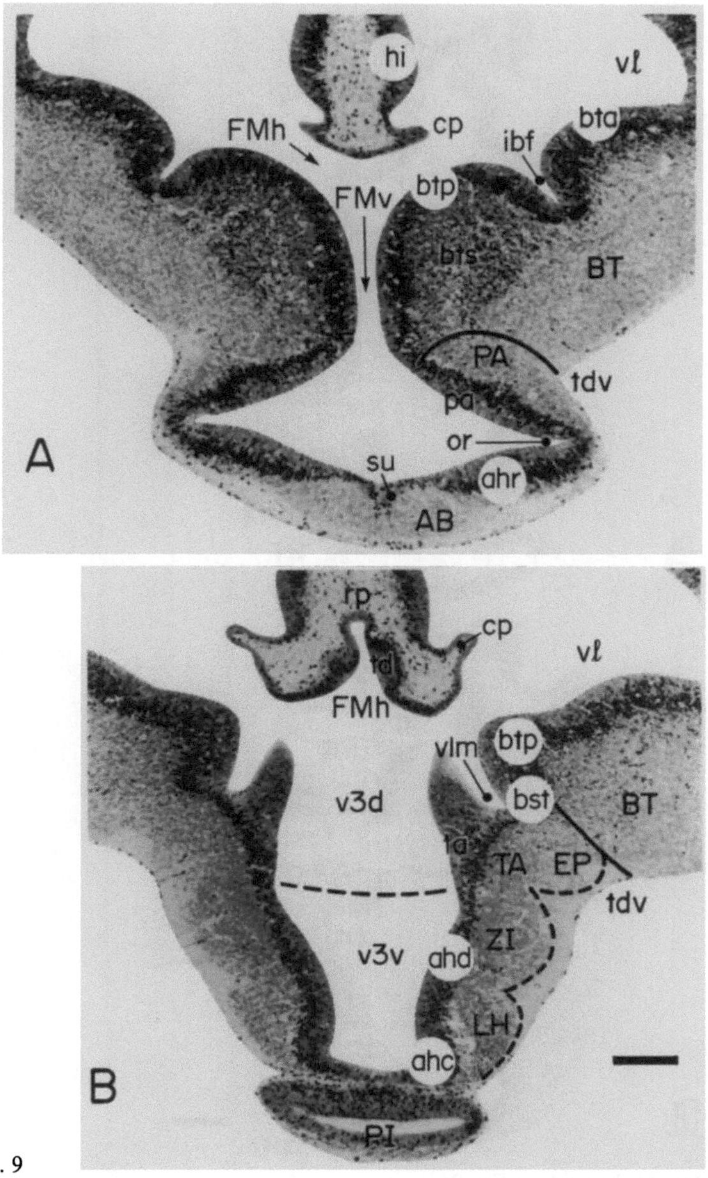

Fig. 10. Legend see Fig. 9

early-settling neurons in the diencephalon form the lateral preoptic area (PA in Fig. 6A) and the lateral mammillary nucleus (ML in Fig. 6E). Settling neurons are not yet evident in the thalamus.

The changes by *day E15* in the configuration of the lumen of the third ventricle and its neuroepithelial lining, and the growth in the differentiating (unlabeled) components of the diencephalon, are shown at 4 sagittal levels (Figs. 7, 8) and in 11 corresponding coronal planes (Figs. 9–13) in radiograms from two rats that received [^{3}H]thymidine at about 9:00 a.m. on this day and were killed 2 h later. One important change is the growth of the thalamic neuroepithelium relative to that of the hypothalamic neuroepithelium. However,

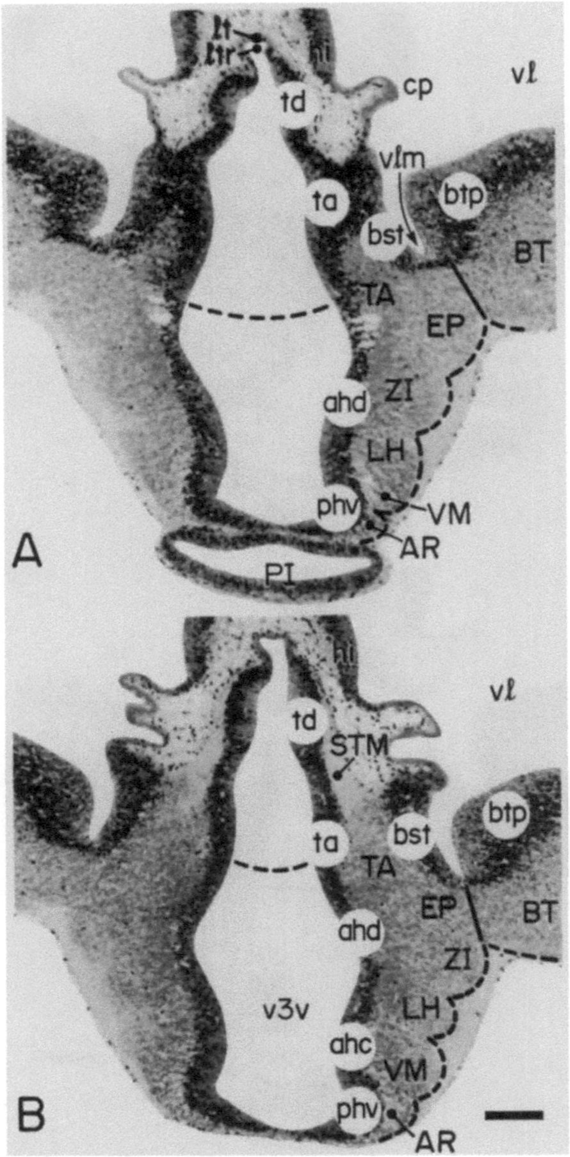

Fig. 11. Legend see Fig. 9

only in two thalamic regions are large clusters of differentiating neurons evident. One of them is the lateral habenular nucleus (HL; Figs. 7B, 9A, B, 13) associated with the habenulopeduncular tract (HP; Figs. 8B, 13B, C); the other is an unidentified early-forming anterior thalamic nucleus (TA; Figs. 8, 10B, 11) described in detail earlier (Altman and Bayer 1979a: Figs. 2, 3, 5), which is associated with the stria medullaris (Figs. 8A, 11B). In contrast, differentiation in the hypothalamus is quite advanced and structures derived from the second wave of migrating neurons are becoming evident.

Rostrally (Fig. 9A, B) the preoptic neuroepithelium is surrounded by the differentiating (unlabeled) cells of the preoptic area (PA). This area is delineated

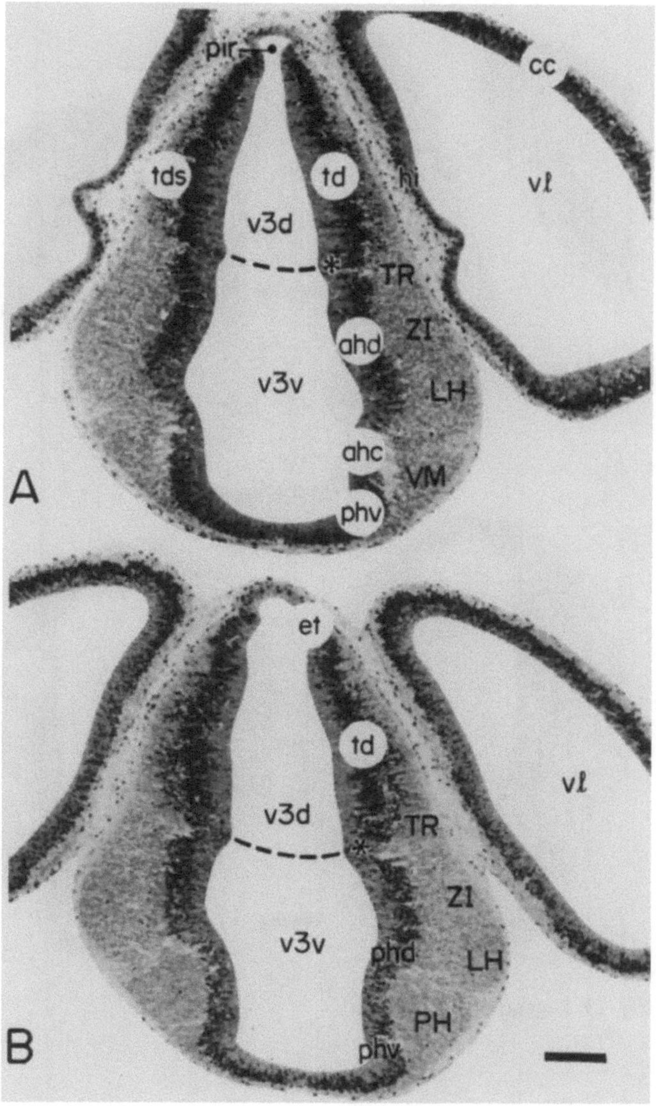

Fig. 12. Legend see Fig. 9

dorsolaterally and dorsally from the telencephalon by an imaginary arc drawn from the ventral telencephalic-diencephalic fissure (tdv) ventrolaterally to the base of the neuroepithelium of the diagonal band of Broca (dB) dorsomedially. A possible separate, wing-like portion of the preoptic area (lateral preoptic area; pal in Fig. 9B) forms the dorsal wall of the optic fissure neuroepithelium. The preoptic area can be traced caudally to the level of the rostral part of the anterior hypothalamic neuroepithelium (ahr in Fig. 10A). At this level the anterobasal nucleus (AB in Figs. 8, 9, 10A), an early-differentiating component of the rostral anterior hypothalamus (Sect. 6.2.1), is prominent. More caudally, the posterior wall of the horizontal branch of the foramen of Monro (FMh in Fig. 10B) separates the rudiments of the dorsal thalamus and the anterior

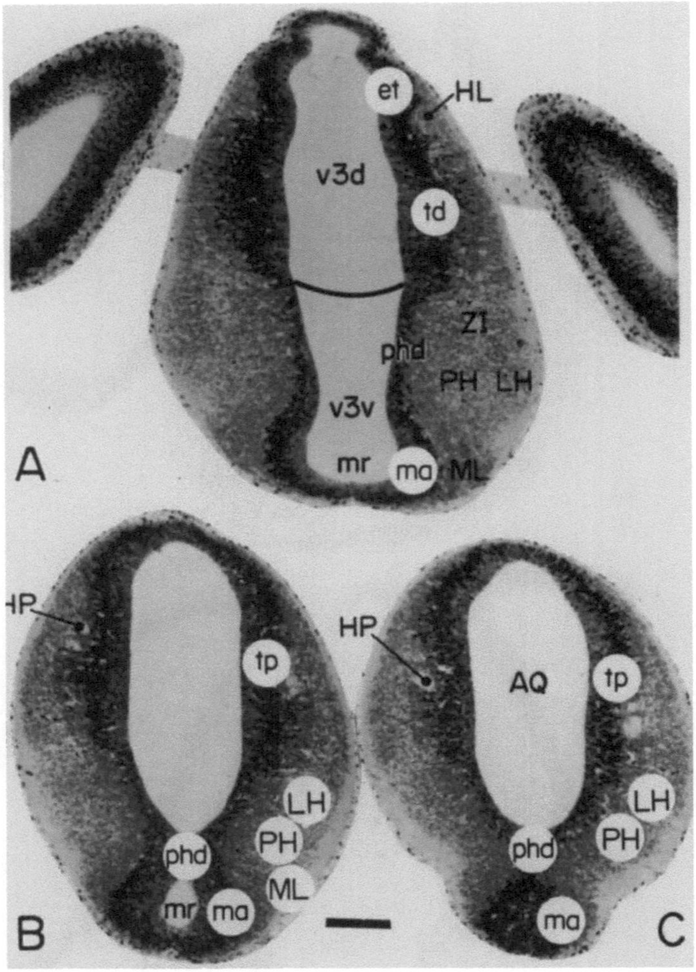

Fig. 13. Legend see Fig. 9

thalamus. Lateral to the anterior thalamic rudiment, flanking the medial horn of the lateral ventricle (vlm in Figs. 10B, 11A), is the neuroepithelium of the bed nucleus of the stria terminalis (bst in Figs. 11B, 12A, B). It is apparently not a derivative of the third ventricle (Sect. 6.2.4). The two major changes with respect to day E14 in the appearance of the bulk of the hypothalamic neuroepithelium are its recession medially and its relative shortening dorsoventrally with respect to the now enlarging thalamic neuroepithelium. Concurrently with the medial translocation of the hypothalamic neuroepithelium, the differentiating zones of the entopeduncular nucleus, zona incerta, and lateral hypothalamus have become greatly enlarged (Figs. 11, 12). Day E15 is the last day of neurogenesis of the entopeduncular nucleus (Bayer 1985: Fig. 4), the zona incerta (Altman and Bayer 1979a Fig. 10), and the lateral hypothalamic area (Altman and Bayer 1978a: Figs. 6, 7), and only a very small proportion of these neurons are produced on this day.

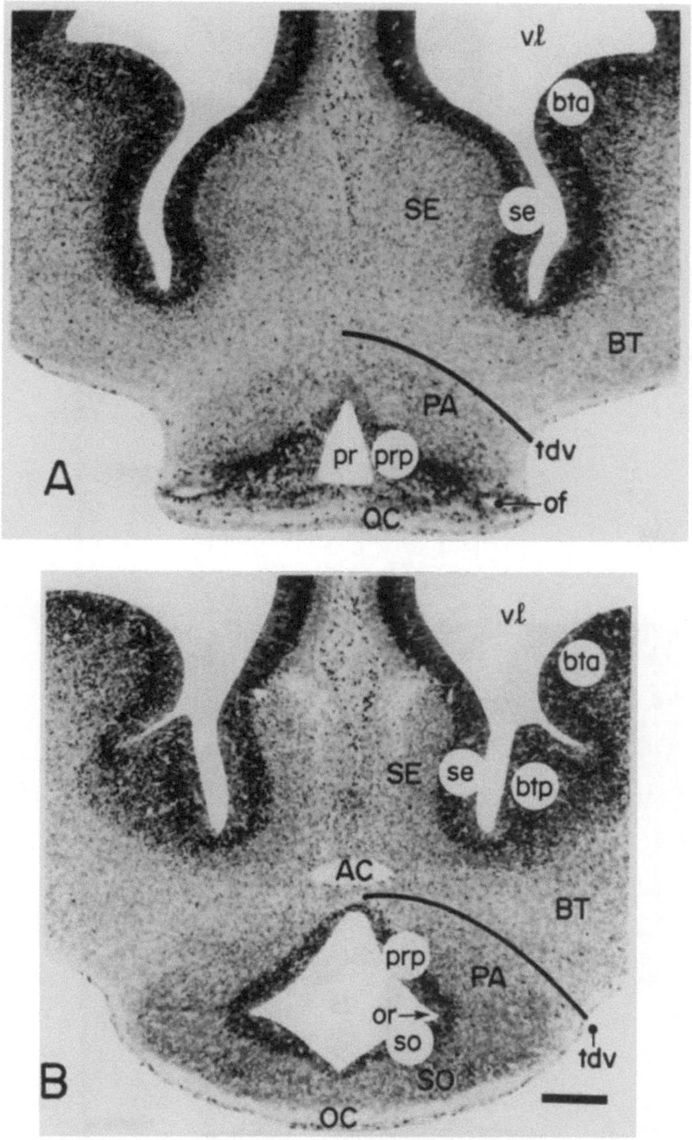

Figs. 14–16. Coronal thymidine radiograms from rostral (Fig. 14 A) to caudal (Fig. 16 B) from a rat labeled on the morning of day E17 and killed 2 h later. Paraffin. *Scales*, 200 μm

Accordingly, the medially displaced hypothalamic neuroepithelum is considered to produce the neurons of the second migratory wave, those contributing to components of the core hypothalamus. Among the structures of the core hypothalamus that are beginning to settle, apart from the preoptic area and the anterobasal nucleus (Figs. 9, 10) referred to earlier, are the ventromedial nucleus (VM in Figs. 11 A, B, 12 A) derived from the caudal portion of the anterior hypothalamic rudiment, the early components of the posterior area nuclei (PH in Figs. 12 B, 13 A–C), and the lateral mammillary nucleus (ML in Fig. 13 A, B).

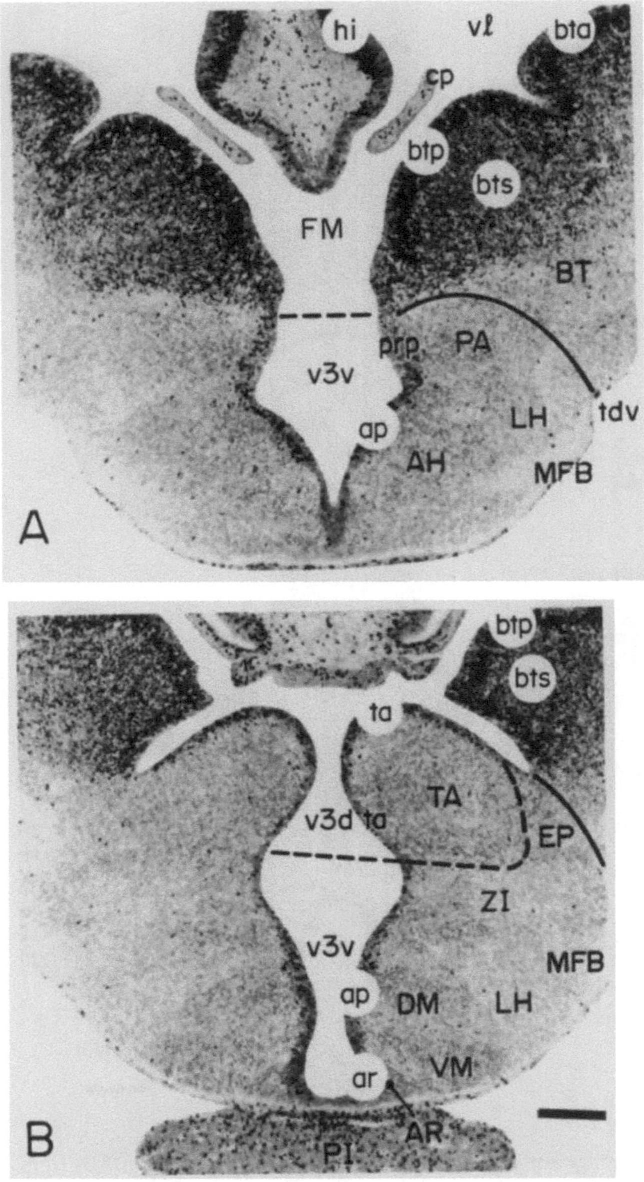

Fig. 15. Legend see Fig. 14

The bulk of the core hypothalamic neurons are generated before day E17 (Altman and Bayer 1978a). The hypothalamic structures which receive an appreciable proportion of neurons on day E17 or thereafter are the preoptic periventricular area, the suprachiasmatic nucleus (Altman and Bayer 1978a: Fig. 5), the anterior periventricular area (not previously examined), the tuberal magnocellular nucleus, and the arcuate nucleus (Altman and Bayer 1978a: Fig. 8). With the exception of the tuberal magnocellular nucleus (Sect. 6.4.2), these are considered to represent the third wave of hypothalamic neurons, those of the midline hypothalamus. The ongoing production of precursors of these neurons is illus-

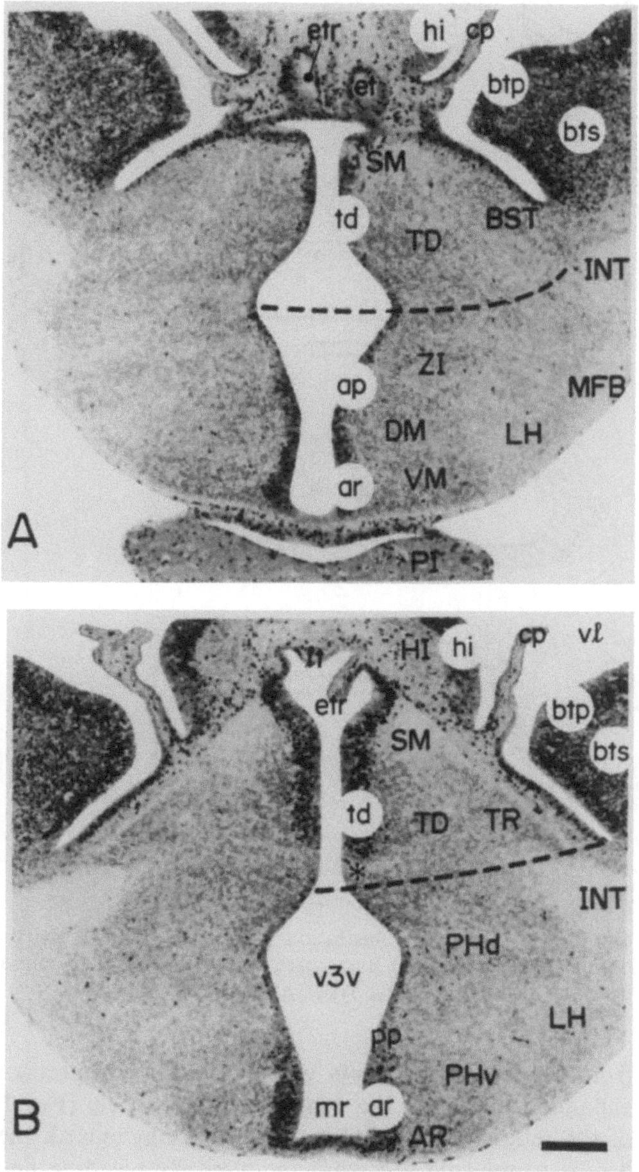

Fig. 16. Legend see Fig. 14

trated in coronal sections from a rat injected on day E17 and killed 2 h later (Figs. 14–16). By this age the settling of the differentiating neurons of the core hypothalamus is considerably advanced. The proliferating cells of the lining of the hypothalamic third ventricle are presumed to be the source also of the specialized lining of the third ventricle ependyma, which are generated principally perinatally (Sect. 8).

Comments. The hypothalamus is divided into three tiers from lateral to medial: the reticular hypothalamus, the core hypothalamus, and the midline hypothalamus. This division is based on the chronology of the production of hypothalamic neurons. The first wave of neurons of the reticular hypothalamus originates

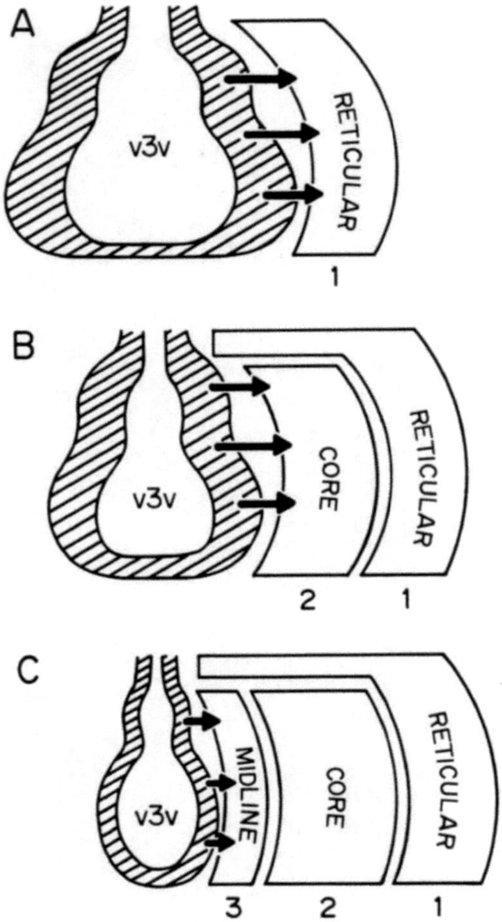

Fig. 17 A–C. Schematic illustration of the derivation of the three tiers of the hypothalamus (reticular, core, midline) from three succeeding waves of migratory neurons (*arrows in* **A–C**) from the medially receding hypothalamic neuroepithelium

from the neuroepithelium while this still occupies a far-lateral position in the expanded lumen of the ventral third ventricle (Fig. 17). Before the generation of the reticular hypothalamic neurons has ceased, the second wave of neurons begins to migrate out from the medially receding neuroepithelium. These neurons form the structures of the core hypothalamus, some of which are becoming recognizable while the third wave of neurons begins to leave the receding and thinning neuroepithelium. The latter form the midline hypothalamic structures which settle in the vicinity of the specialized lining of the hypothalamic third ventricle. The core hypothalamus is also divided into four major longitudinal zones (Fig. 2) on the basis of conspicuous landmarks in the early-developing third ventricle (the preoptic recess, the optic fissure, the infundibular recess, and the mammillary recess). These are from rostral to caudal: the preoptic area, the anterior hypothalamus, the posterior hypothalamus, and the mammillary body. Further subdivisions are the rostral and caudal anterior hypothalamic regions, and the ventral and dorsal posterior hypothalamic regions (which are divided during early development by the mammillary recess).

5 Development of the Reticular Hypothalamus: The First Wave

The term "reticular" is used by anatomists to designate a netlike brain region where the cell bodies of neurons are scattered among criss-crossing fiber bundles. The term has also acquired a physiological meaning when the mesencephalic reticular formation has been implicated in arousal and wakefulness (Magoun 1950). The term "reticular" is used here strictly in the anatomical sense to designate three components of the hypothalamus which constitute an ontogenetically uniform system and which, by coincidence or not, are characterized in adults by a reticular organization. As shown in Fig. 18, the differentiating neurons of the entopeduncular nucleus (EP in Fig. 18A), the zona incerta (ZI in Fig. 18B, C), and the lateral hypothalamic area (LH in Fig. 18A–C) are settling as early as day E14, while the neuroepithelium is still situated far-laterally around the bulging lumen of the third ventricle. The presumptive sites of origin of these structures (ep, zi, and lh) are indicated.

5.1 The Lateral Hypothalamus

Background. The lateral hypothalamic area has received considerable attention recently because of its involvement in a variety of behavioral functions, such as feeding (Brügger 1943; Anand and Brobeck 1951), predatory aggression (Wasman and Flynn 1962; King and Hoebel 1968), and sensory neglect and akinesia (Marshall and Teitelbaum 1974). These multifarious functions may at least partially be related to the circumstance that this diffuse region, which extends from the lateral preoptic area rostrally to the level of the mammillary body caudally, is traversed throughout its extent by a heterogeneous collection of ascending and descending fibers, the medial forebrain bundle (Nieuwenhuys et al. 1982). By way of the medial forebrain bundle the lateral hypothalamus is interconnected rostrally, among other structures, with the medial prefrontal cortex, the septum, the nucleus accumbens, the amygdala, and the bed nucleus of the stria terminalis, and caudally with the mesencephalic tegmentum and central gray, and with certain pontine and medullary nuclei (Saper et al. 1979; Shiosaka et al. 1980; Kita and Oomura 1982). Although the exact synaptic relationship of these mostly unmyelinated fibers with the lateral hypothalamic neurons are not known, collaterals terminating here have been described (Millhouse 1969).

The embryonic development of the lateral hypothalamus has received little attention. According to Ifft (1972), heavily labeled lateral hypothalamic neurons are seen in the rat after injection of [^{3}H]thymidine between days E12 and E16, with the highest concentration on day E14. In our study with the cumulative

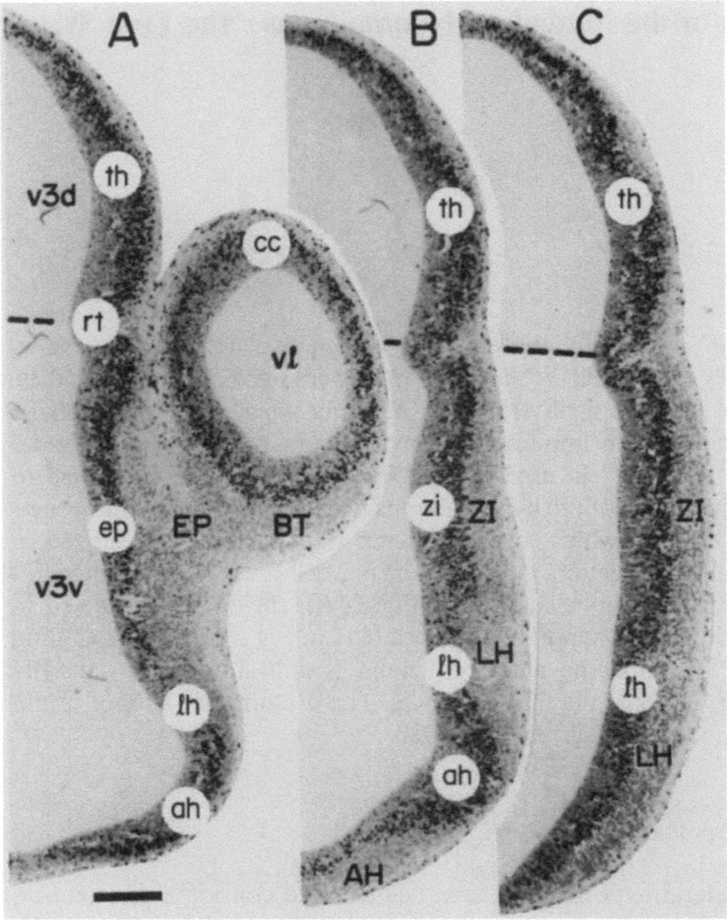

Fig. 18A–C. Midcoronal thymidine radiograms of the third ventricle neuroepithelium from rostral (**A**) to caudal (**C**) from a rat labeled on the morning of day E14 and killed 2 h later. At these levels the reticular protuberance (*rt*) divides the hypothalamic from the thalamic neuroepithelium. The settling of the neurons of the entopeduncular nucleus, zona incerta, and lateral hypothalamus is evident at this early age. Methacrylate. *Scale*, 200 μm

labeling procedure (Altman and Bayer 1978a: Figs. 6, 7), we have found that these neurons are generated between days E12 and E15, with peak production time on day E13. There was no appreciable difference in this pattern between rostral and caudal levels.

Results. In perinatal rats that received [^{3}H]thymidine on day E13, practically all the lateral hypothalamic (LH) neurons are labeled (Fig. 19A). Within the area, the cells situated far-laterally (enclosed region in Fig. 20A) tend to be heavily labeled; those situated medially and ventromedially (arrows in Fig. 20A) are mostly lightly labeled. In rats that received [^{3}H]thymidine on day E14 (Fig. 19B), the heavily labeled cells are still numerous, but these tend to be concentrated more medially and ventromedially, in a concentric ring (arrows in Fig. 20B) around the far-lateral portion (enclosed region in Fig. 20B), where a high proportion of the cells are no longer labeled. These observations indicate

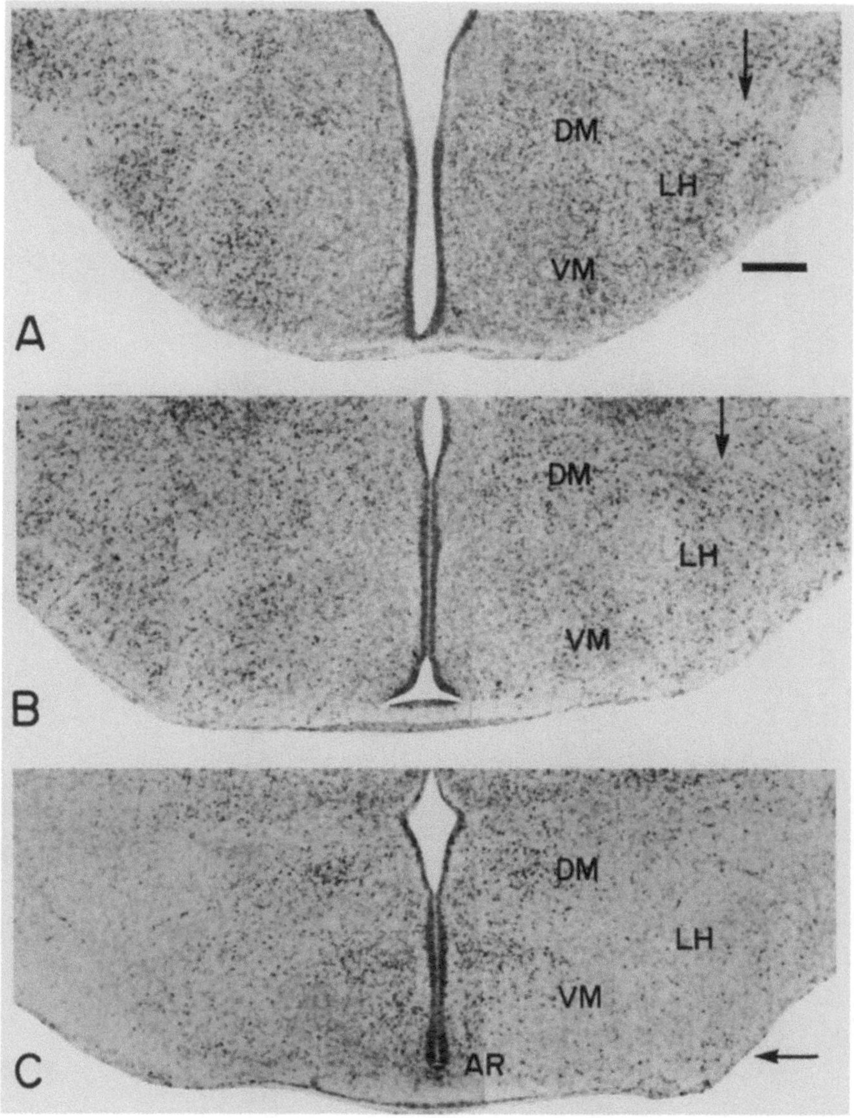

Fig. 19 A–C. Coronal thymidine radiograms from perinatal rats labeled on day E13 (**A**), E14 (**B**) and E15 (**C**). In the lateral hypothalamus heavily labeled cells are seen far-laterally following injection on day E13 (*arrow in* **A**) and mid-laterally following injection on day E14 (*arrow in* **B**). Few heavily labeled cells are present in the rat injected on day E15, except ventrally (*horizontal arrow in* **C**). Paraffin. *Scale*, 200 μm

a weak lateral-to-medial neurogenetic gradient in the lateral hypothalamus. No such gradient was observed along the rostrocaudal axis. In rats that received [^{3}H]thymidine on day E15 (Fig. 19C) most of the lateral hypothalamic neurons were no longer labeled. Some of the heavily labeled cells were scattered throughout the area; others were concentrated ventromedially (arrow in Fig. 19C).

We have made an attempt to locate the site of origin of neurons of the lateral hypothalamic area in radiograms from rats that received [^{3}H]thymidine

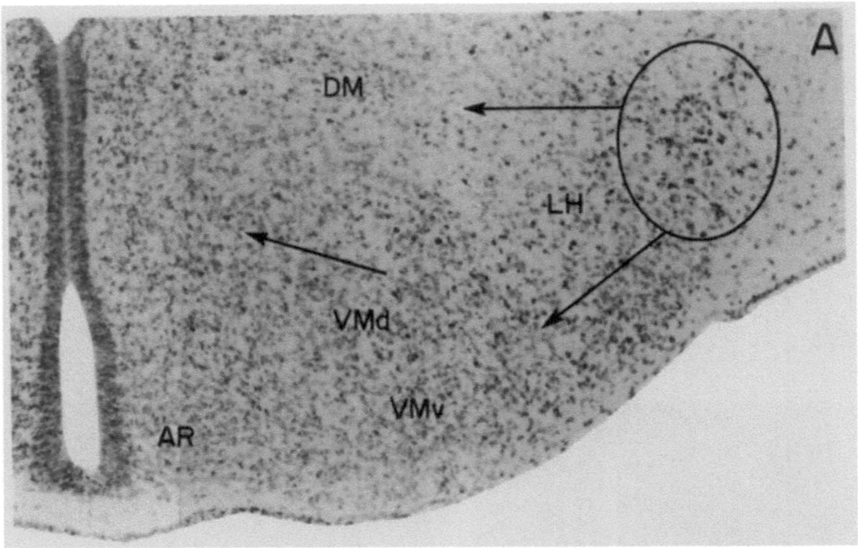

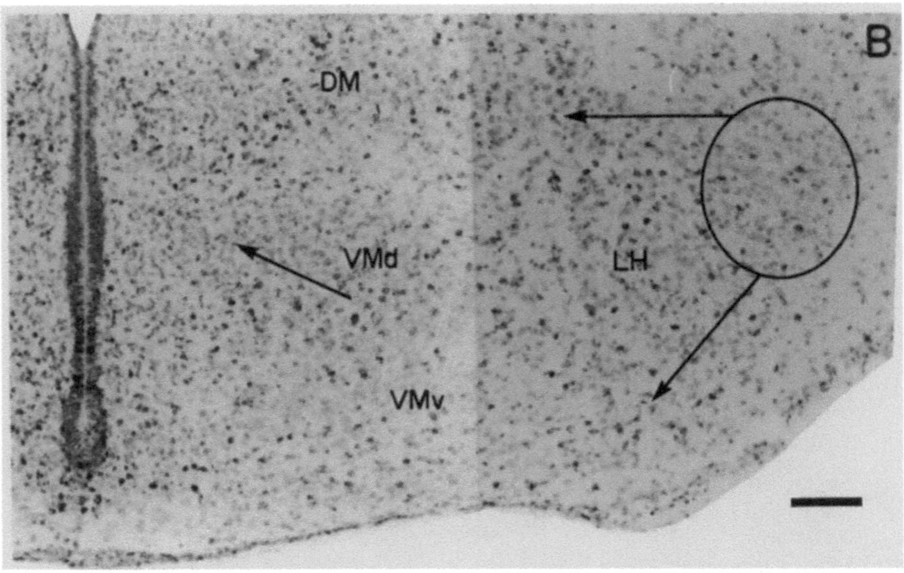

Fig. 20 A, B. Coronal thymidine radiograms at higher magnification to show the labeling pattern in the lateral hypothalamus in perinatal rats labeled on day E13 (**A**) and day E14 (**B**). The neurogenetic gradients (*arrows*) in the lateral hypothalamus and in the ventromedial nucleus are indicated. Paraffin. *Scale*, 100 μm

on day E14 and were killed sequentially thereafter. In rats that were killed 2 h after injection (Fig. 21 A), the neuroepithelium that surrounds the wide lumen of the third ventricle is flanked laterally by a narrow zone of unlabeled cells. These may be lateral hypothalamic neurons that were generated on day E13 or earlier. In rats that were killed 1 day after injection (Fig. 21 B), the neuroepithelium has receded medially to fill in the lumen of the third ventricle, and the lateral hypothalamic area has expanded considerably. However, a large

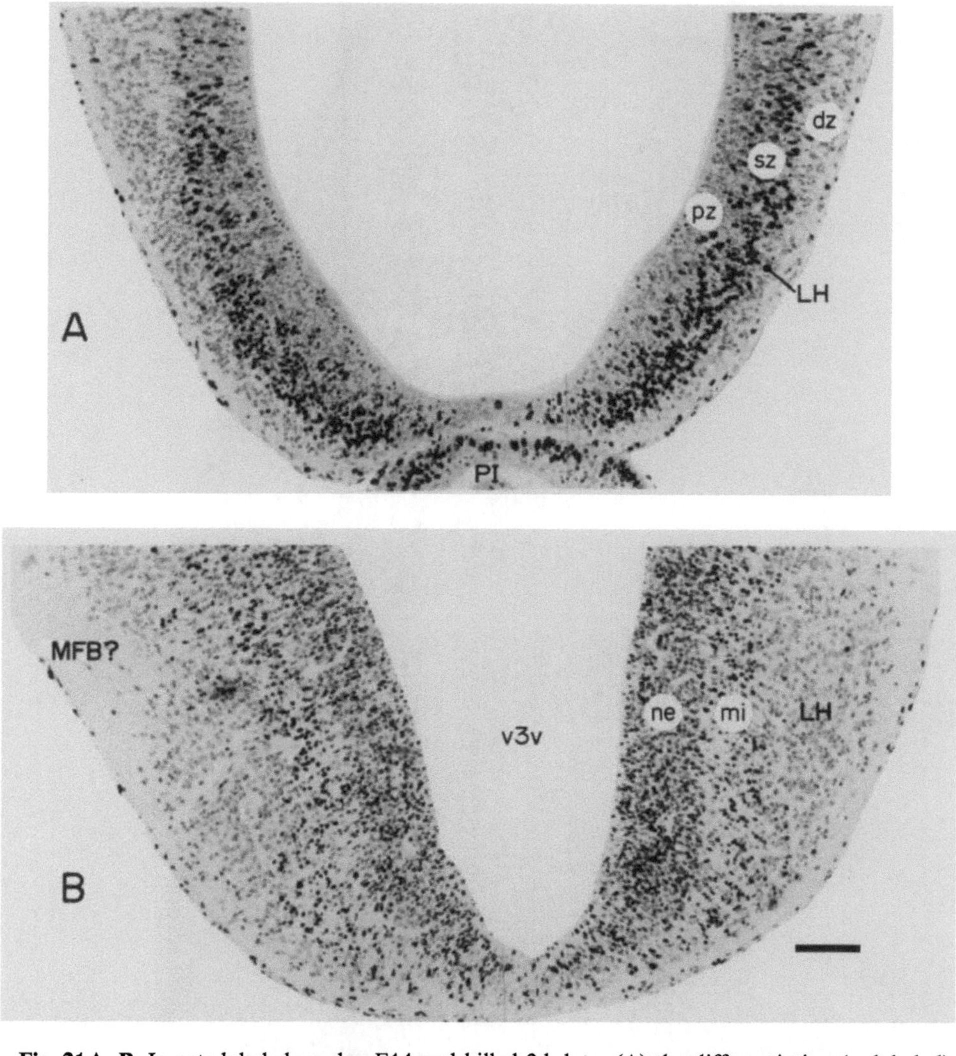

Fig. 21 A, B. In rats labeled on day E14 and killed 2 h later (**A**) the differentiating (unlabeled) zone of the lateral hypothalamus is relatively narrow but it becomes broader in rats injected on the same day and killed 1 day later (**B**). This suggests nonradial migration from adjacent areas. Radial migration is suggested by the band of heavily labeled cells surrounding the neuroepithelium in **B**. Methacrylate. *Scale*, 100 μm

complement of these newly added cells are unlabeled. These cells, therefore, must have been generated before day E14. Apparently, some cells generated on day E13 or earlier were still migrating in from some more distant site. The heavily labeled cells around the neuroepithelium (mi in Fig. 21 B) may be radially migrating cells or are destined to reach adjacent areas. The white matter above the growing lateral hypothalamic area may be composed of the approaching fibers of the medial forebrain bundle (MFB in Fig. 21 B). In rats killed 2 days after injection (day E16; Fig. 22 A), labeled cells begin to penetrate the greatly expanded lateral hypothalamic area and these cells are now embedded in a matrix presumably composed of fibers of the medial forebrain bundle.

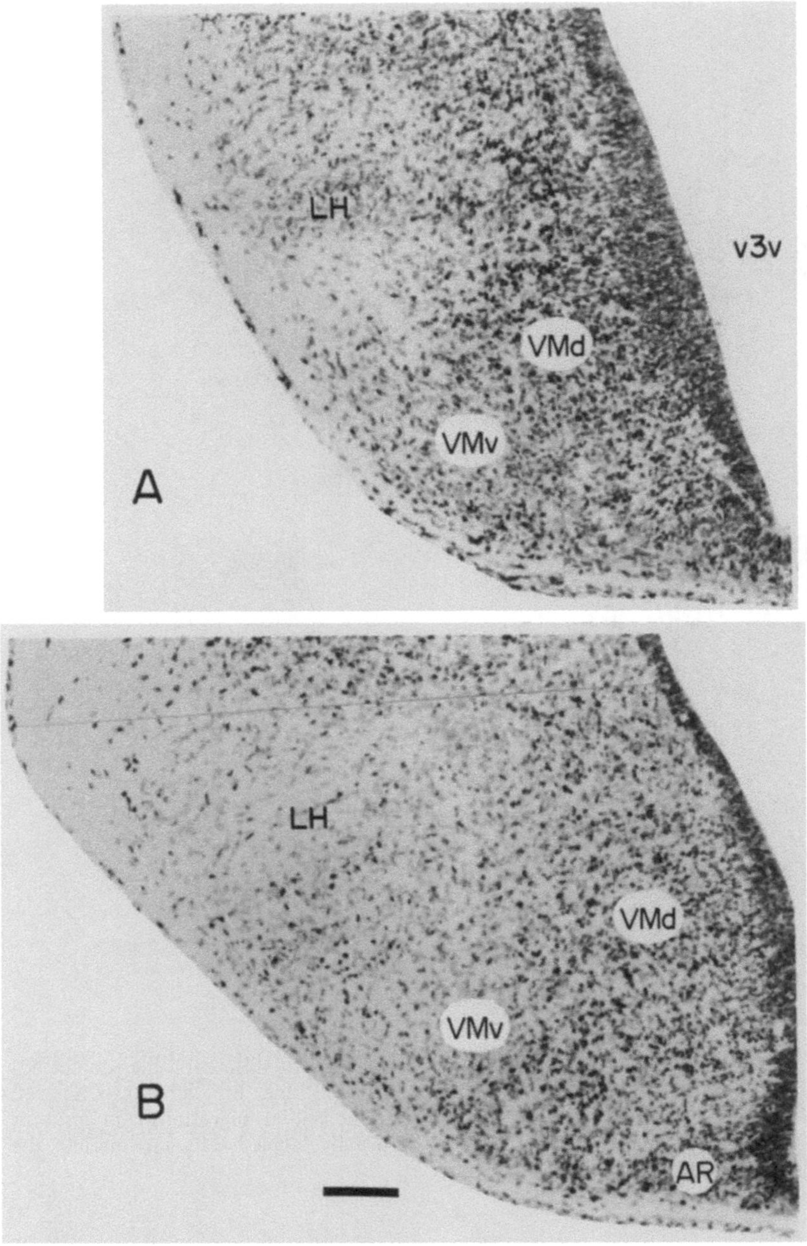

Fig. 22A, B. The progressive dispersion of the neurons of the lateral hypothalamus, presumably associated with the penetration of fibers of the medial forebrain bundle, is illustrated in thymidine radiograms from rats injected on day E14 and killed on day E16 (**A**) or day E17 (**B**). Paraffin. *Scale*, 100 μm

This process continues on the next day (E17; Fig. 22B), and the dispersal of labeled and unlabeled cells in the greatly enlarged lateral hypothalamic area begins to resemble the maturing pattern (Fig. 19B).

The possible site of production of lateral hypothalamic neurons in shown in horizontal radiograms from rats injected on day E14 and killed sequentially.

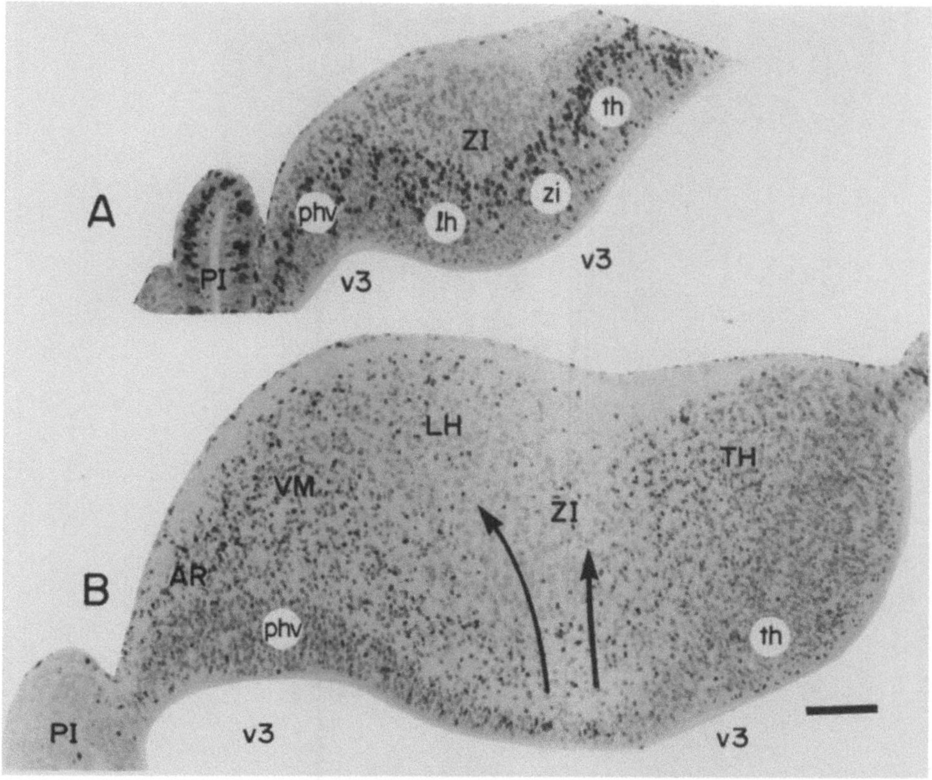

Fig. 23 A, B. Horizontal thymidine radiograms from rats labeled on day E14 and killed 2 h (**A**) or 2 days (**B**) later. The putative lateral hypothalamic neuroepithelium is thick and mitotically active in the day E14 rat but is largely exhausted in the day E16 rats (*source of curved arrow in* **B**). The putative germinal source of the zona incerta is also indicated. Methacrylate. *Scale*, 100 μm

In rats killed 2 h after injection (Fig. 23 A) a relatively large differentiating zone is present with unlabeled cells posterior to the thalamic region (which, as yet, is devoid of a differentiating zone). The broader anterior component is presumed to be the early-differentiating cells of the zona incerta (ZI) and the narrower posterior component the early-differentiating cells of the lateral hypothalamic area. The neuroepithelium is broad, and highly active mitotically along the entire length of the third ventricle at this stage of development. In rats killed 2 days after injection (age E16; Fig. 23 B), when lateral hypothalamic neurogenesis has come to an end, there is a thin neuroepithelium (ventricular lining) interposed between the still large thalamic (th) and ventral posterior hypothalamic (phv) neuroepithelia. We tentatively assume that the neurons of lateral hypothalamus and the zona incerta originated from this region on the preceding days. Finally, the presumptive site of origin of the neurons of the lateral hypothalamus is illustrated in coronal sections of day E17 rats that received [^{3}H]thymidine either on day E15 (Fig. 24 A), the last day of lateral hypothalamic neurogenesis, or on day E16 (Fig. 24 B), after hypothalamic neurogenesis has come to an end. The lateral hypothalamic neuroepithelium is distinguished from other

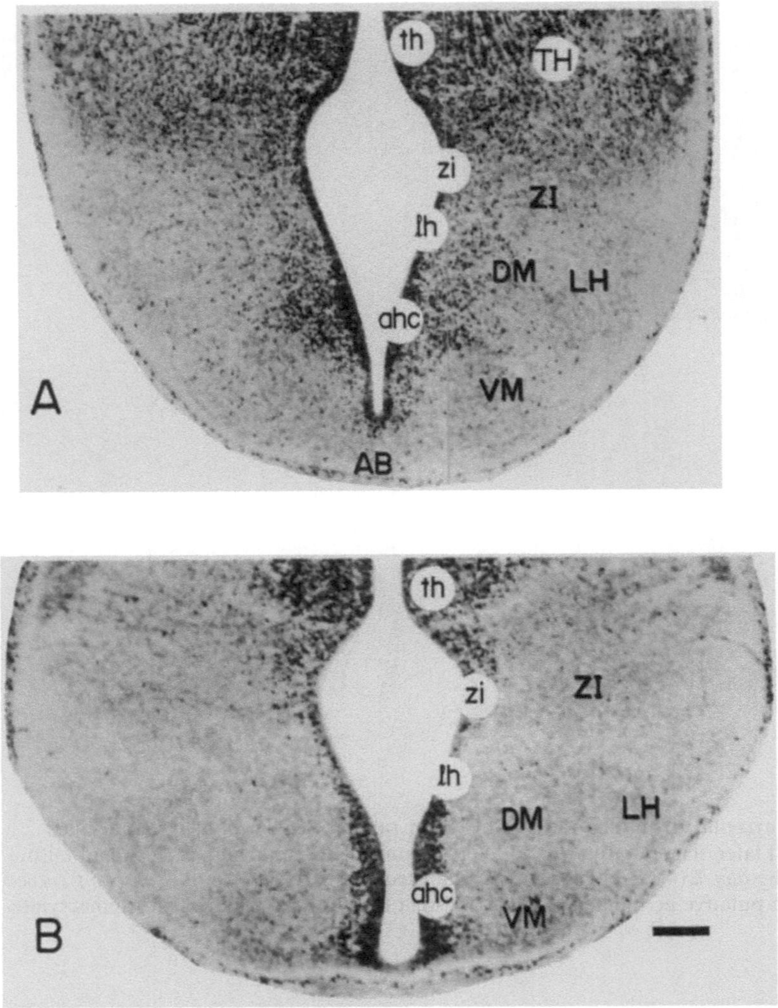

Fig. 24A, B. The exhausted neuroepithelial zones of the lateral hypothalamus and the zona incerta are indicated in coronal radiograms from rats injected on day E15 and killed on day E17 (**A**) and injected on day E16 and killed on day E17 (**B**). Paraffin. *Scale*, 200 μm

segments of the lining of the third ventricle, the neuroepithelium of the zona incerta excepted, by its thinness and progressive exhaustion of labeled cells.

Comments. These observations are in agreement with our previous quantitative results (Altman and Bayer 1978a: Figs. 6, 7) that the lateral hypothalamic area neurons are among the earliest-produced cells of the hypothalamus and that their generation comes to an end on day E15. Observations in sequential embryonic radiograms suggest that these neurons arise from a segment of the third ventricle neuroepithelium that is situated caudally beneath the neuroepithelium of the zona incerta and rostrally beneath the neuroepithelium of the entopeduncular nucleus (Figs. 23, 24). The lateral hypothalamic area is, accordingly, associated ontogenetically with components of the zona incerta and the entopeduncular nucleus, which many anatomists have classified with the ventral thalamus and the telencephalon (basal ganglia), respectively. Our indirect observations, based on the penetration of white matter and the dispersal of the neurons of the lateral hypothalamic area (Figs. 21B, 22A), suggest that the fibers of the medial forebrain bundle do not become distributed in the area until after day E15, after the settling of a fair proportion of the lateral hypothalamic neurons. But this needs to be examined with specific fiber-staining techniques. Unlike so many other hypothalamic structures, the lateral hypothalamic area is diffusely organized from the beginning of its embryonic maturation.

5.2 The Entopeduncular Nucleus

Background. In the rat, the large neurons of the entopeduncular nucleus are embedded in the posteroventral portion of the internal capsule. The entopeduncular nucleus is generally considered to be the subprimate homologue of the medial portion of the globus pallidus (Fox et al. 1966). This homology is supported by the similarity of the afferent supply and efferent outflow of these two structures. The major afferent supply of the medial pallidum in the monkey is from the striatum (Voneida 1960; Szabo 1962), and the same applies to the entopeduncular nucleus in the cat (Voneida 1960) and the rat (Van der Kooy and Carter 1981). In the monkey, the efferent outflow of the medial pallidum is to the ventral lateral, ventral anterior, and centromedian nuclei of the thalamus, to the lateral habenular nucleus, and to the nucleus tegmenti

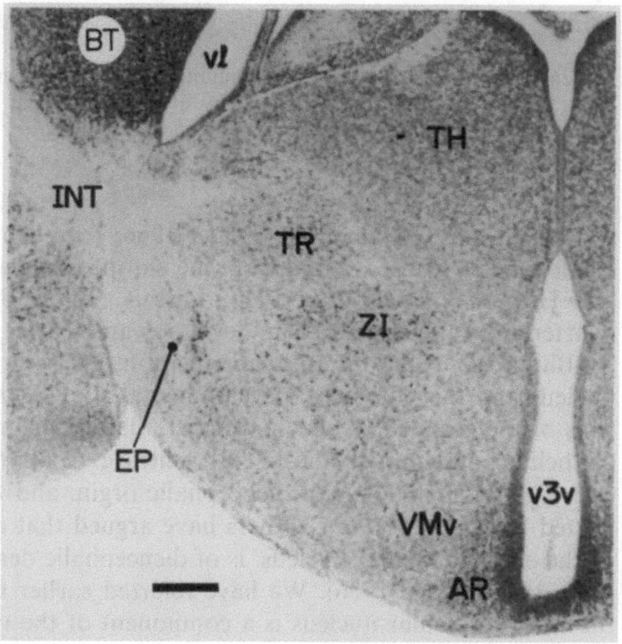

Fig. 25. Low-power coronal thymidine radiogram showing the heavily labeled (and some unlabeled) cells of the entopeduncular nucleus in a perinatal rat injected on day E13. Many labeled cells in the zona incerta are also seen. Paraffin. *Scale*, 200 μm

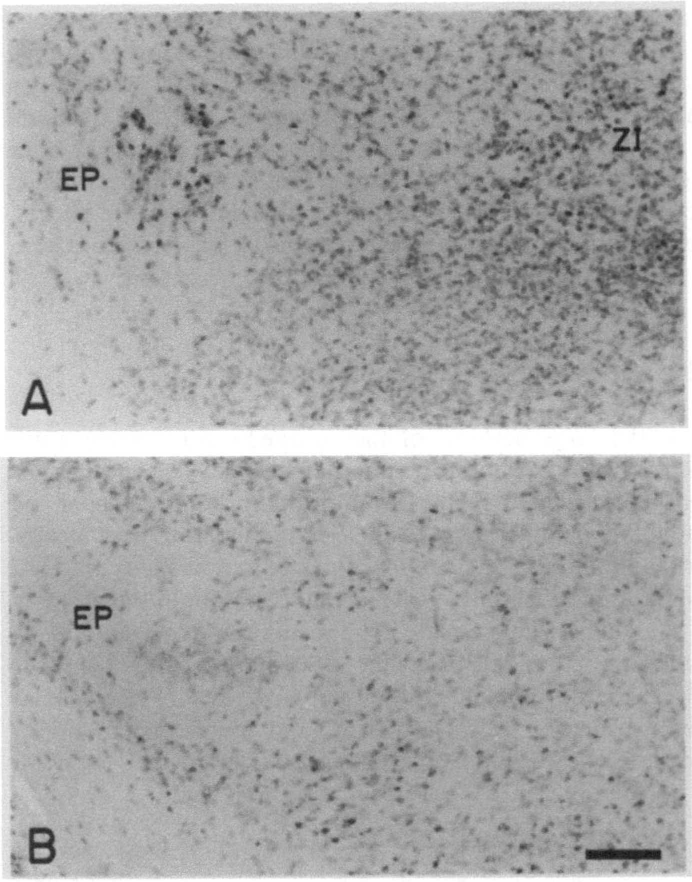

Fig. 26 A, B. High-power coronal thymidine radiograms from perinatal rats tagged on day E13 (**A**) and day E14 (**B**). The majority of entopeduncular neurons are heavily labeled in **A**, but are no longer labeled in **B**. Paraffin. *Scale*, 100 µm

pedunculopontis (Nauta and Mehler 1966; Kim et al. 1976). The same efferent outflow has also been found from the entopeduncular nucleus in the cat (Filion and Harnois 1978; Nauta 1979; Larsen and McBride 1979) and in the rat (Carter and Fibiger 1978; Van der Kooy and Carter 1981).

But in spite of this intimate relationship with the striatum (the caudate nucleus and the putamen), developmental studies have indicated that the neurons of the entopeduncular nucleus are derived from the third ventricle neuroepithelium, not that of the lateral ventricle. Spatz (1925) maintained that the entire globus pallidus is of diencephalic orgin, and this view was recently supported by Richter (1965). Others have argued that only the external pallidum, or the entopeduncular nucleus, is of diencephalic derivation (Kuhlenbeck 1923; Rose 1942; Kahle 1956). We have referred earlier to the controversy whether the entopeduncular nucleus is a component of the ventral thalamus or the hypothalamus; the latter view has been advocated by Rose (1942) and Kuhlenbeck (1977). In a recent quantitative thymidine radiographic study (Bayer 1985), it was established that the neurons of the rat entopeduncular nucleus are gener-

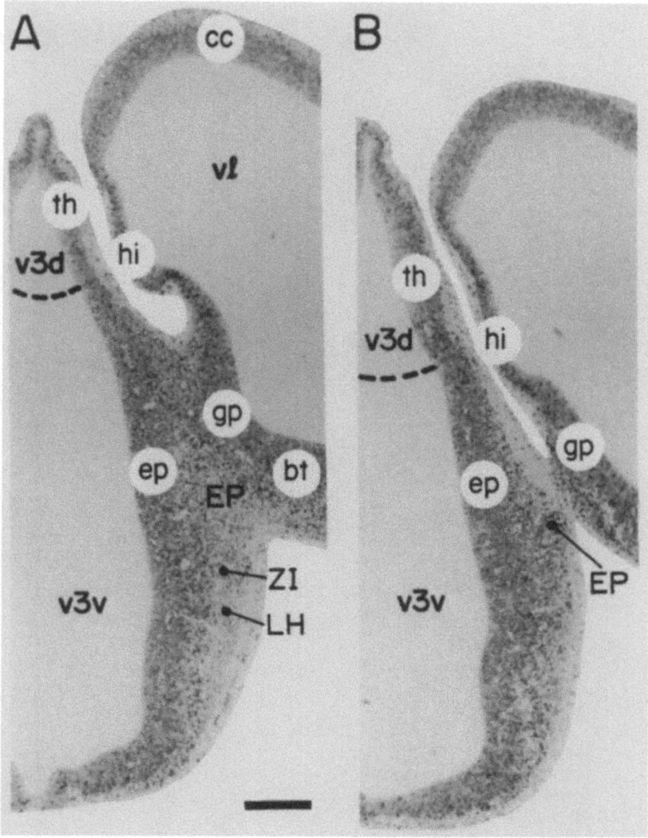

Fig. 27 A, B. Coronal thymidine radiograms from rostral (**A**) to caudal (**B**) from a rat labeled on day E13 and killed on day E14. The diencephalic origin of the heavily labeled entopeduncular neurons is evident caudally. Methacrylate. *Scale*, 200 μm

ated principally on days E12 and E13 with a "sandwich" pattern of neurogenetic gradient.

Results. In perinatal rats that received [^{3}H]thymidine on day E13, the early-produced entopeduncular neurons (EP in Figs. 25, 26A) stand out with their intense labeling (and some of its unlabeled cells) against a background of lightly labeled cells throughout the rest of the diencephalon (with the exception of the similarly early-produced neurons of the zona incerta). In rats that were tagged with [^{3}H]thymidine on day E14, the majority of the entopeduncular neurons were no longer labeled (Figs. 26B, 32). In rats that received [^{3}H]thymidine on day E13 and were killed 1 day later, the entopeduncular nucleus, with an admixture of heavily labeled and unlabeled cells, is already recognizable (Figs. 27, 28), indicating an early outflow of the differentiating neurons. Anteriorly, these cells are situated between the apposed neuroepithelia of the third and lateral ventricles (Fig. 27A) and, therefore, their site of origin is not obvious. However, most posteriorly it is evident that the entopeduncular neurons derive from the third ventricle (Figs. 27B, 28). The apposed neuroepithelium of the

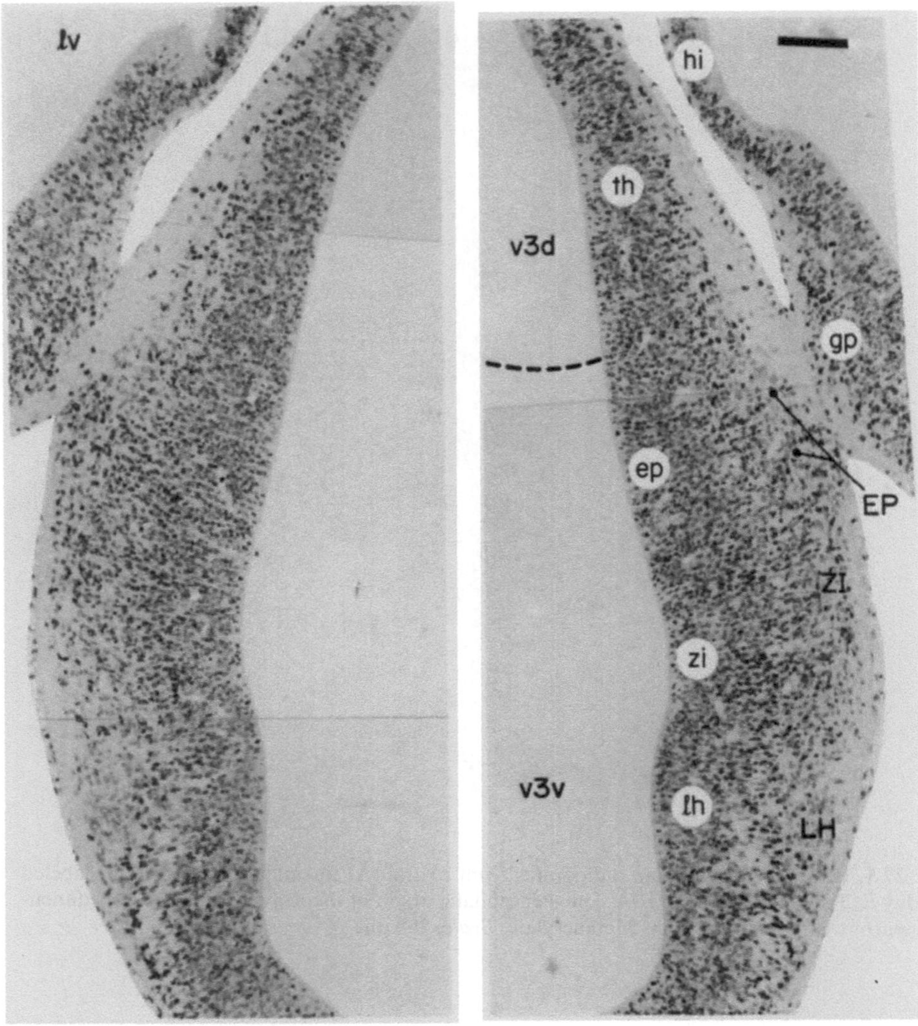

Fig. 28. Illustration of the presumptive diencephalic location of the entopeduncular neuroepithelium dorsal to that of the zona incerta in a rat injected on day E13 and killed on day E14. Methacrylate. *Scale*, 100 μm

basal telencephalon is presumed to generate the neurons of the external globus pallidus (gp in Figs. 27, 28, 63). In rats injected on day E13 and killed 2 days afterwards (Figs. 29, 30) the later-produced (lightly labeled) pallidal neurons (Bayer 1985) can be seen settling external to the earlier-produced (unlabeled and heavily labeled) entopeduncular neurons.

Comments. These observations confirm that the entopeduncular nucleus is composed of very early produced neurons. They also support the view of Kuhlenbeck (1977) and Rose (1942) that the nucleus derives from the dorsal portion of the hypothalamic neuroepithelium. However, the view of Spatz (1925) and Richter (1965) that the external globus pallidus is also of diencephalic origin

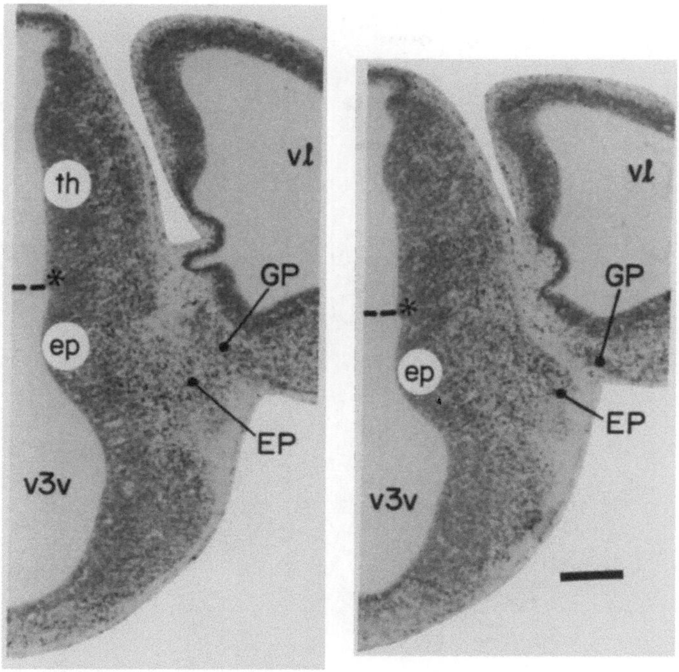

Fig. 29 A, B. Low-power coronal thymidine radiograms from rostral (**A**) to caudal (**B**) from a rat injected on day E13 and killed on day E15 to show the apparent diencephalic origin of the entopeduncular nucleus and telencephalic origin of the globus pallidus. Paraffin. *Scale*, 200 μm

is not supported. It appears, rather, that the two components of the globus pallidus are of different derivation: the internal portion (the entopeduncular nucleus) being diencephalic, and the external portion telencephalic. The entopeduncular neurons appear to be settling between the junction of the telencephalon and diencephalon quite early (Fig. 27), much before the possible penetration of this region by thalamocortical or corticothalamic fibers.

5.3 The Zona Incerta

Background. The zona incerta spreads beneath the thalamus almost along the entire length of the diencephalon. At anterior levels it is contiguous dorsolaterally with the thalamic reticular nucleus, and posteriorly with the ventral nucleus of the lateral geniculate body. The zona incerta is composed of scattered cells traversed by a heterogeneous collection of fibers. Ricardo (1981) reviewed the reports dealing with afferents to the zona incerta. These include fibers from the spinal cord, the nuclei gracilis and cuneatus, the principal sensory trigeminal nucleus, the cerebellum, the deep layers of the superior colliculus, the mesencephalic reticular formation, the periaqueductal gray, the pretectum, the ventromedial nucleus of the hypothalamus, and several areas of the cerebral cortex. Kaelber and Smith (1979) distinguished two components of the zona incerta

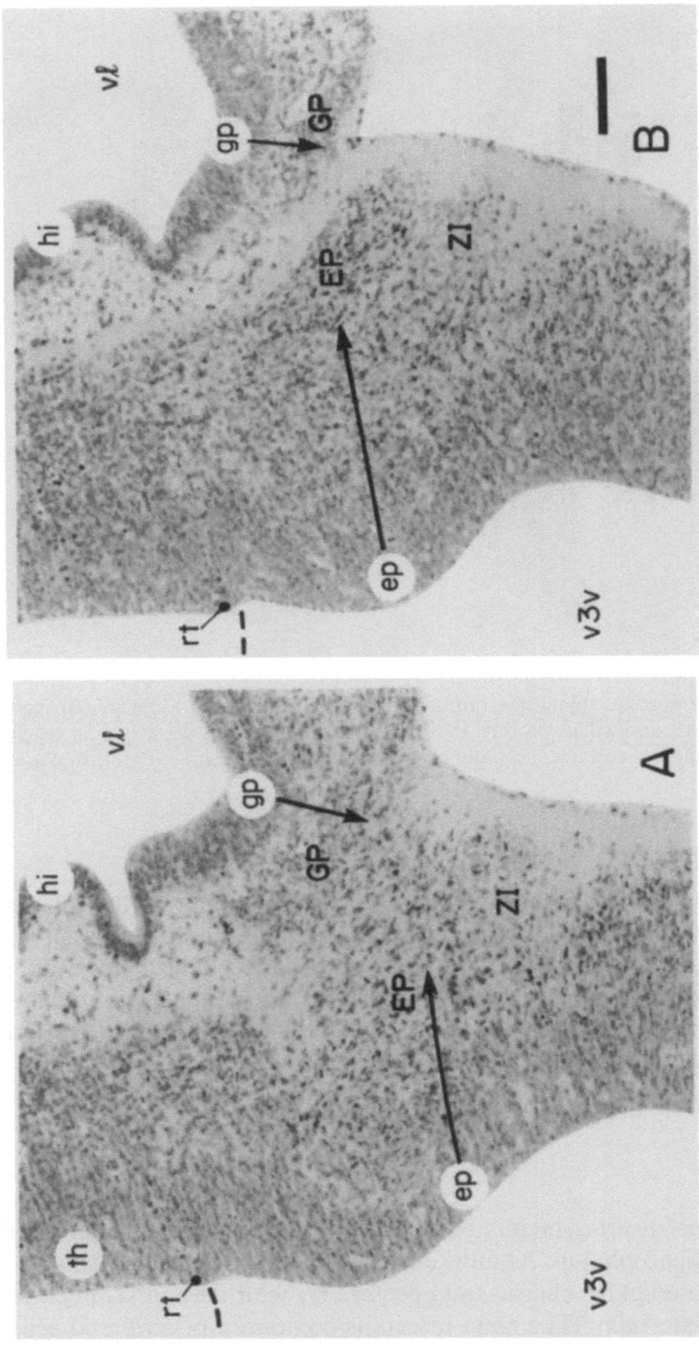

Fig. 30 A, B. Detail of Fig. 29. *Arrows* indicate the presumed route of the heavily labeled entopeduncular neurons and lightly labeled globus pallidus neurons from their presumptive diencephalic and telencephalic sites of origin. Paraffin. *Scale*, 100 μm

in the cat, a medial zone implicated in nociceptive functions and a lateral zone implicated in extrapyramidal motor functions. The medial zone projects, among other structures, to the intralaminar nuclei of the thalamus, portions of the hypothalamus, and the mesencephalic central gray. The lateral zone projects to the entopeduncular nucleus and to all the components of the striatum. An

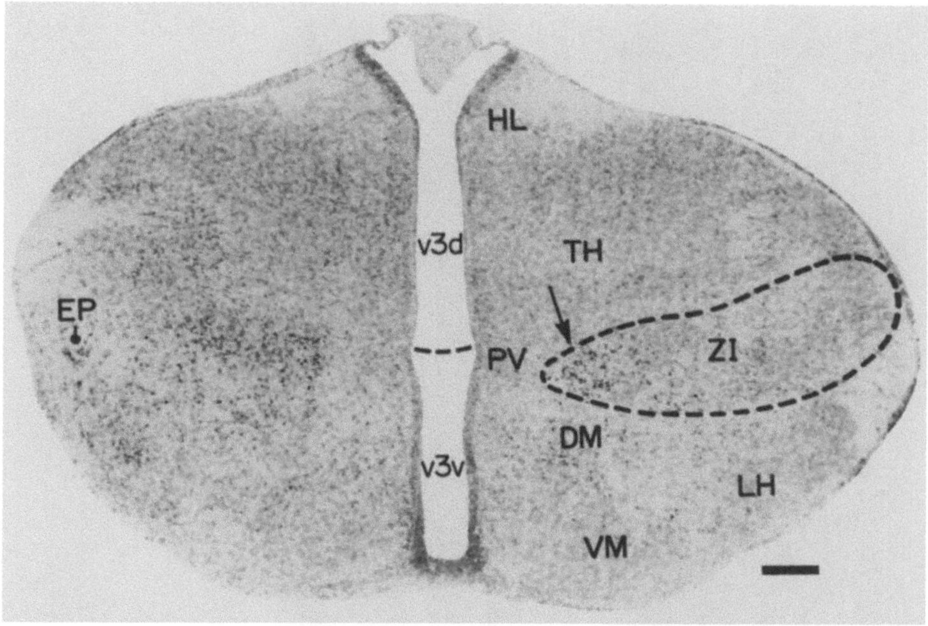

Fig. 31. Thymidine radiogram from a perinatal rat labeled on day E13 to show the medial concentration of the heavily labeled (earliest) cells of the zona incerta. Paraffin. *Scale*, 200 μm

even more extensive efferent outflow from the zona incerta has been found by Ricardo (1981) in the rat. This includes, in addition to the above, several mesencephalic and pontine reticular structures, the precerebellar nuclei, the peri-oculomotor nuclei, and the red nucleus. A projection to the hippocampal formation has also been reported in the rat (Köhler et al. 1984). Thus, the connections of the zona incerta appear to be as widespread as that of the lateral hypothalamus. Little attention has been paid in the past to the development of this diffuse region. Our quantitative studies have shown that the neurons of the zona incerta are generated between days E13 and E15, with a peak on day E13 (Altman and Bayer 1979a: Fig. 10).

Results. In perinatal rats that received [^{3}H]thymidine on day E13, the medial part of the zona incerta (arrow in Fig. 31) is distinguished by its collection of heavily labeled cells; the cells in the rest of the zona incerta (broken outline) are lightly labeled. In rats tagged on day E14, there is a core of unlabeled cells in the medial part of the zona incerta (ZIm in Fig. 32) which tapers off laterally (ZIl). This core of unlabeled cells, which are generated before the morning of day E14, is surrounded by a sheath of heavily labeled cells. In rats tagged on day E15, labeled cells are limited in the zona incerta to its far-lateral part (ZIf in Fig. 33). Accordingly, the bulk of the zona incerta cells covering its core must be generated before the morning of day E15.

The site of origin and settling pattern of the zona incerta neurons are illustrated in rats that received [^{3}H]thymidine on day E14 and were killed on days E15 (Fig. 34A), E16 (Fig. 34B), and E17 (Fig. 34C). Just beneath the reticular

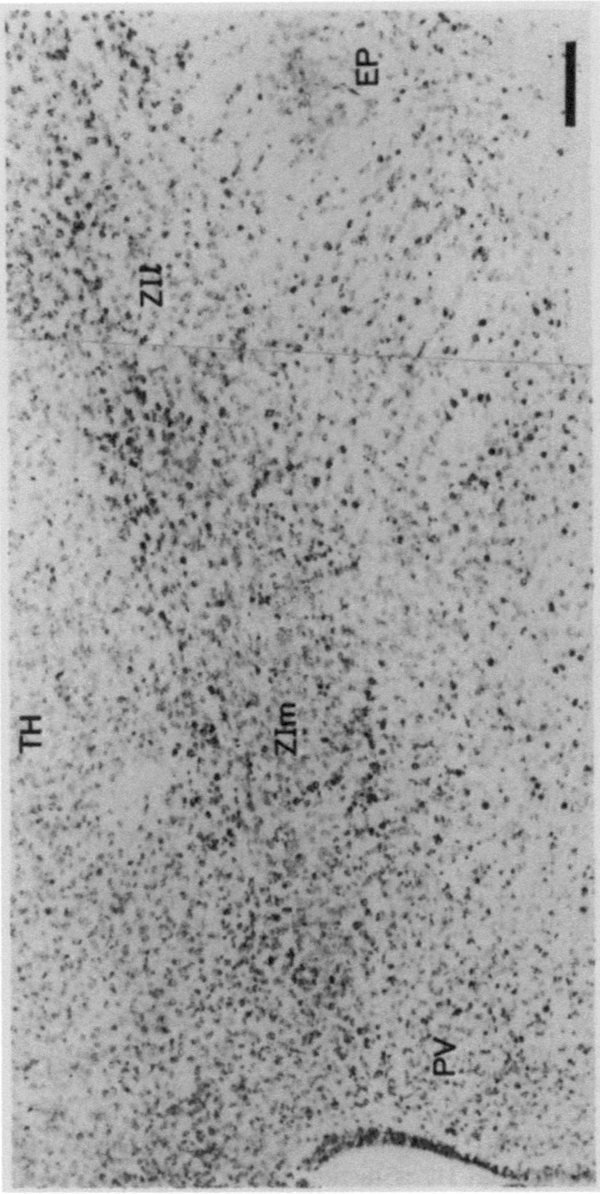

Fig. 32. In perinatal rats labeled on day E14 there is a core of unlabeled cells medially and midlaterally. Paraffin. *Scale*, 100 µm

protuberance (asterisk in Fig. 34A), an evaginated neuroepithelial lobule is flanked by a differentiating mass, composed of unlabeled and labeled cells. The former is identified as the neuroepithelium of the zona incerta, the latter as its early-settling neurons. This neuroepithelial zone has been identified earlier as a part of the rostral region of the anterior hypothalamic rudiment. By day E16 (Fig. 34B), the zona incerta neuroepithelium has decreased in width and the zona incerta is greatly enlarged by extending both medially (toward the shrinking third ventricle) and laterally. By day E17 (Fig. 34C), the three parts

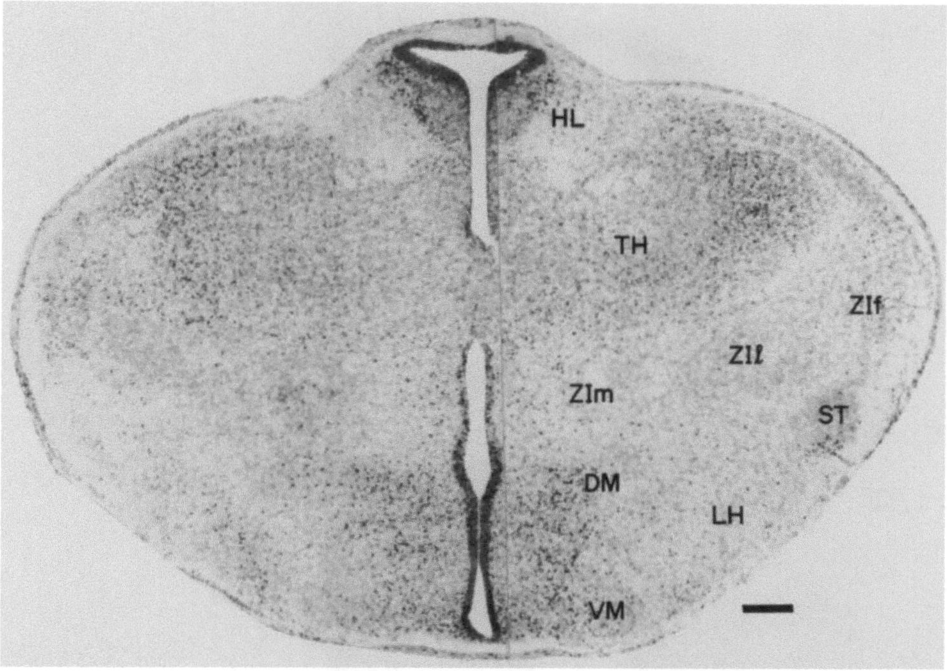

Fig. 33. In perinatal rats labeled on day E15, labeled cells in the zona incerta are limited to its far-lateral region. Paraffin. *Scale*, 200 μm

of the zona incerta are recognizable. The concentration of labeled cells tends to be higher laterally than medially. The outflow pattern of the last-generated neurons of the zona incerta is illustrated in rats that were tagged on day E15 and killed on day E16 (Fig. 35A) and day E18 (Fig. 35B). One day after injection a small but compact migratory stream of labeled cells appears to be moving outward (arrow in Fig. 35A) in the direction of the earlier-settled (unlabeled) neurons of the lateral part of the zona incerta (ZIl in Fig. 35A). By day E18 the labeled cells (those produced after the morning of day E15) appear to have settled in the far-lateral part of the zona incerta (ZIf in Fig. 35B).

Comments. The present observations are in agreement with our earlier quantitative results that the zona incerta neurons are produced between days E13 and E15 (Altman and Bayer 1979a: Fig. 10). But apparently the production of neurons on day E14 is more extensive than the quantitative data have indicated. The zona incerta neurons originate in the rostral part of the anterior hypothalamic rudiment, dorsal to the neuroepithelium of the lateral hypothalamic area. The neurogenetic gradient of the zona incerta is unusual for the diencephalon, as the medially situated neurons are produced earlier than its far-lateral neurons (inside-out instead of the usual outside-in pattern). Conceivably, the parts of the zona incerta distinguished by different times of origin represent different structures. The different anatomical connections and physiological functions of the medial and lateral (far-lateral?) parts of the zona incerta (Kaelber and Smith 1979) have been referred to earlier.

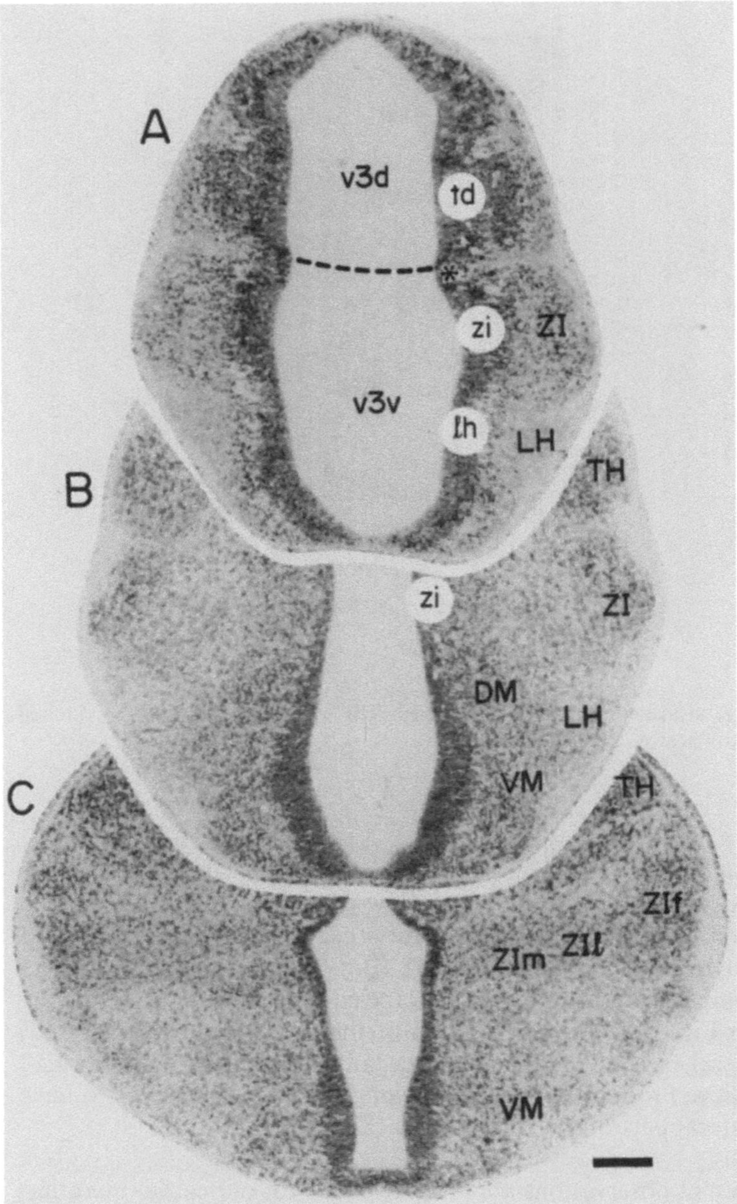

Fig. 34 A–C. Coronal thymidine radiograms from rats injected on day E14 and killed on day E15 (**A**), E16 (**B**) and E17 (**C**) to show the presumed site of origin of the neurons of the zona incerta and their settling in an inside-out sequence. Paraffin. *Scale*, 200 μm

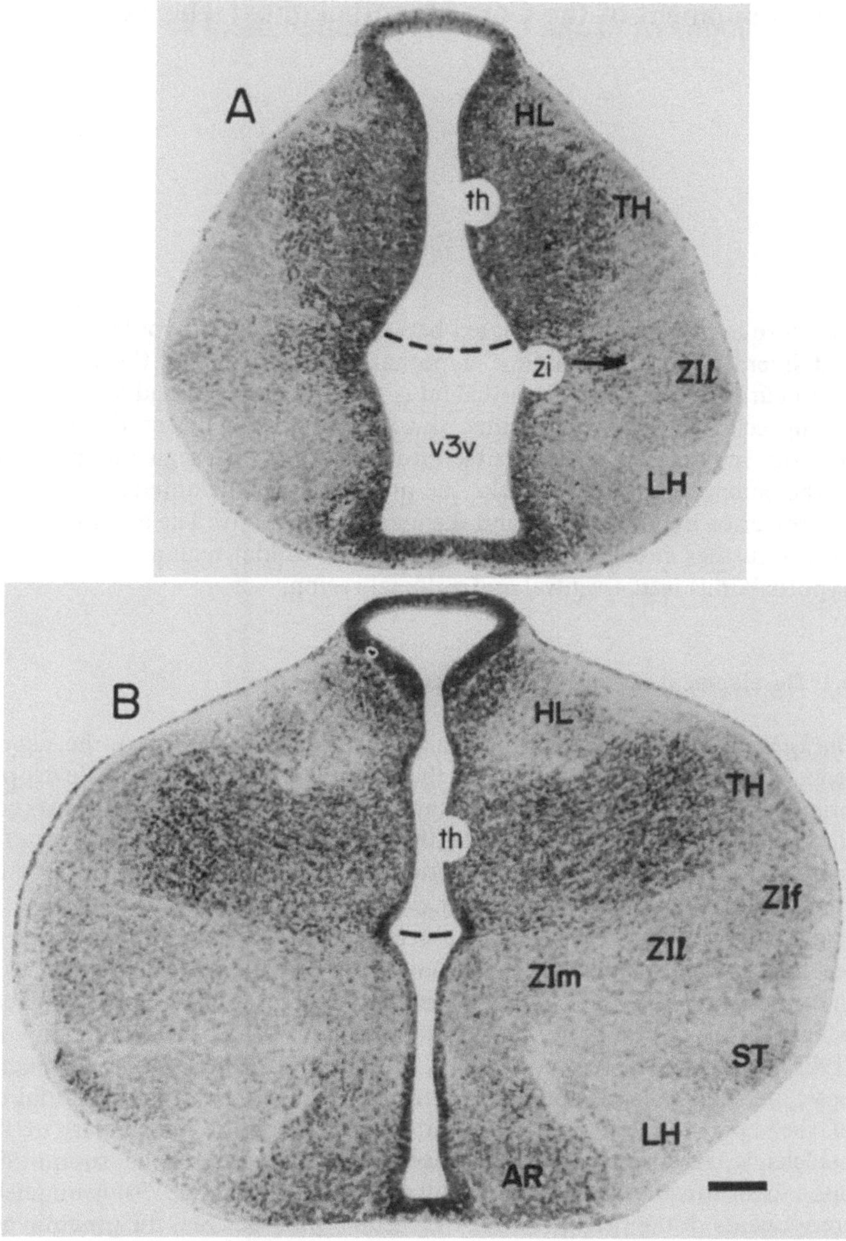

Fig. 35 A, B. Coronal thymidine radiograms from rats labeled on day E15 and killed on day E16 (**A**) and day E18 (**B**) to show the outflow pattern (**A**) and settling (**B**) of the last complement of zona incerta neurons. Paraffin. *Scale*, 200 μm

6 Development of the Core Hypothalamus: The Second Wave

The core hypothalamus is situated between the reticular hypothalamus dorsally and laterally, and the midline hypothalamus adjacent to the third ventricle. The neurons of the core hypothalamus constitute the second wave of differentiating cells of the hypothalamic third ventricle. As it was described earlier, the core hypothalamus is divisible into four transverse segments by landmarks of the embryonic third ventricle: the preoptic area, the anterior hypothalamus, the posterior hypothalamus, and the mammillary body. The anterior hypothalamus is further subdivided into a rostral and caudal region, and the posterior hypothalamus into a ventral and a dorsal region.

6.1 Development of the Preoptic Area

Background. The term preoptic area came into use by the end of the last century (e.g., Edinger 1896) but to this day there is no agreement as to its developmental origins, boundaries, or constituent subdivisions. While most anatomists consider the preoptic area to be a part of the hypothalamus, because its caudal part merges imperceptibly into the anterior hypothalamus, and because of its presumed diencephalic derivation (reviewed by Christ 1969), others have argued (e.g., Le Gros Clark 1938) that it is a part of the telencephalon. One reason for this disagreement may be the uncertainty of the rostral boundaries of the preoptic area, and the circumstance that structures have been assigned to it which are likely to be of telencephalic origin. Thus, according to Gurdjian (1927), the preoptic region includes, in addition to the lateral and medial preoptic areas and the preoptic periventricular nucleus, also the interstitial nucleus of the septal portion of the medial forebrain bundle, and a part of the bed nucleus of the stria terminalis. A recent trend has been to incorporate the preoptic area into the anterior hypothalamus. Bleier et al. (1979) distinguish three components of the preoptic area, the medial preoptic area, the preoptic periventricular nucleus, and the medial preoptic nucleus. In particular, they consider the medial preoptic nucleus to be a continuation of the anterior hypothalamic nucleus and distinguish the two merely by an arbitrary landmark, the former being situated over the rostral half of the optic chiasm, the latter over its caudal half.

Results. The preoptic neuroepithelium (pa in Figs. 3B, 4, 7–9, 10A) is defined as the germinal matrix of the third ventricle that extends initially from the anterior and anterodorsal wall of the optic fissure (pal in Fig. 9B) laterally and to the preoptic recess anteromedially (pr in Figs. 3B, 4A, 5A, 7, 8, 9B).

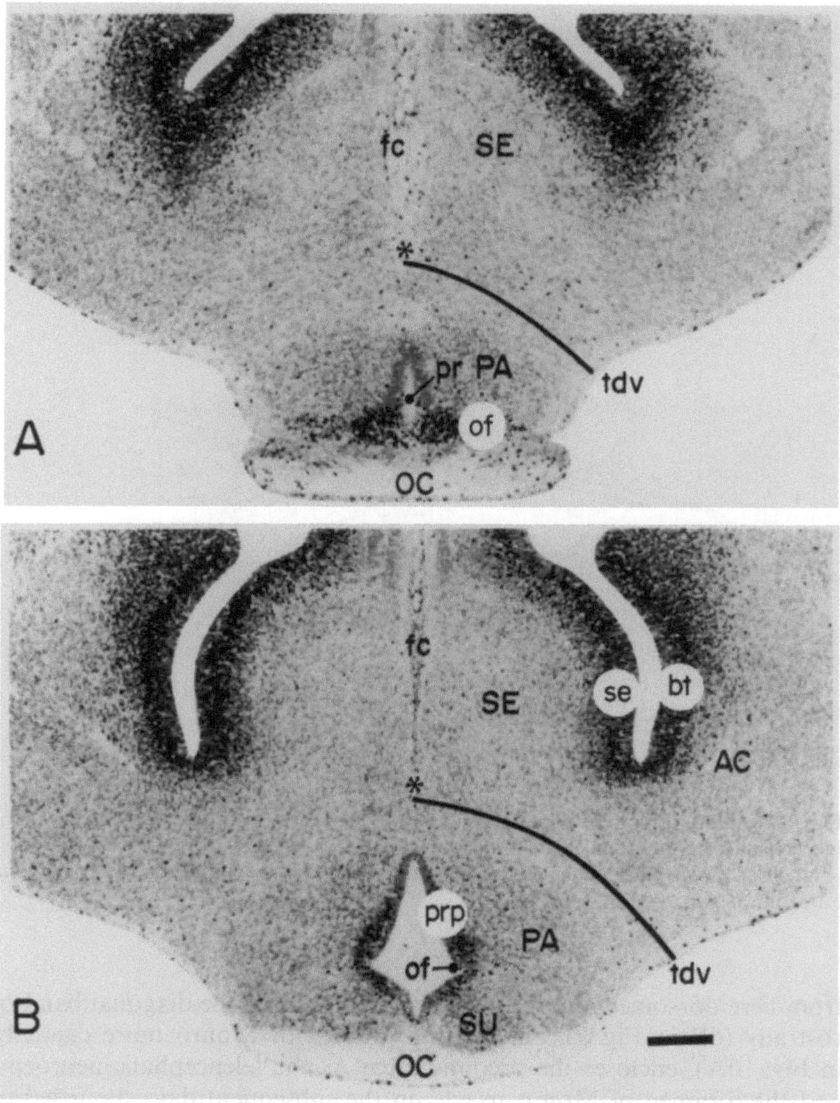

Fig. 36 A, B. Coronal thymidine radiograms from a rat labeled on day E18 and killed 2 h later, from rostral (**A**) to caudal (**B**), showing the optic recess and its reduced but regionally still active neuroepithelium. The preoptic area is delineated by a curve from the base of the falx cerebri (*asterisk*) dorsomedially to the telencephalic-diencephalic fissure ventrolaterally. Paraffin. *Scale*, 200 μm

Later (days E16–E18) the lateral component disappears and the preoptic neuroepithelium lengthens appreciably in the rostral direction (Figs. 14, 36). Structures derived from this portion of the neuroepithelium (that is, the rostral wall of the optic fissure) are considered to be components of the preoptic area (PA in Fig. 36). These will be distinguished from structures derived from the caudal wall of the optic fissure neuroepithelium as components of the anterior hypothalamic area. The ventrolateral boundary of the preoptic area is the ventral telencephalic-diencephalic fissure (tdv in Figs. 9, 10 A, 36). An imaginary arc drawn

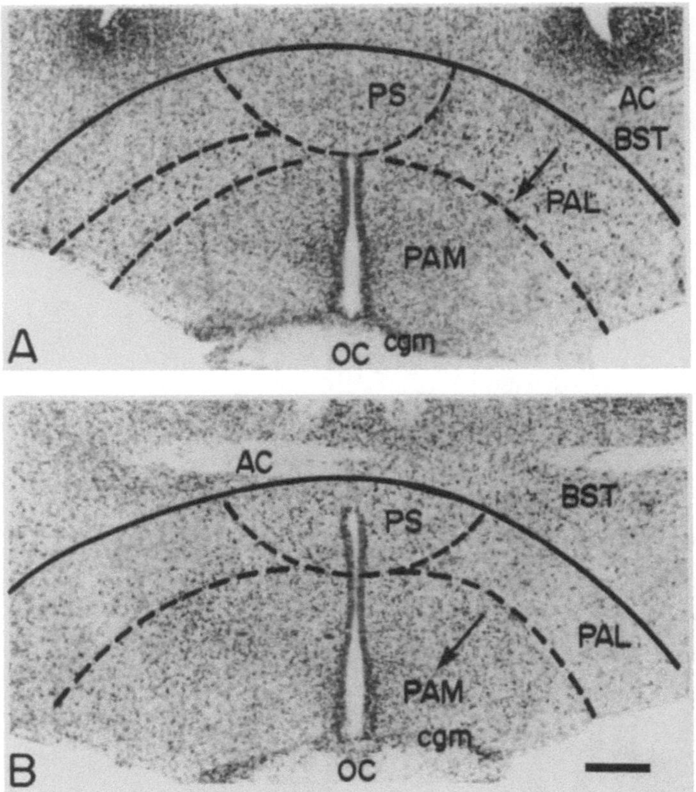

Figs. 37 and 38. Coronal thymidine radiograms from the rostral portion of the preoptic area in perinatal rats injected on days E13 (Fig. 37A), E14 (Fig. 37B), E15 (Fig. 38A), and E16 (Fig. 38B). *Solid line* delineates the rostral preoptic area and *broken lines* its components. Paraffin. *Scales*, 200 μm

from here dorsomedially to the neuroepithelium of the diagonal band of Broca rostrally (dB in Fig. 9), and to the foramen of Monro more caudally (FMv in Fig. 10A), encloses the preoptic area. As the telencephalic neuroepithelium and the foramen of Monro recede on the subsequent days the new middorsal landmark caudally is the base of the septum, and the anterior commissure (Figs. 14, 36–44), which apparently crosses the midline through the interface of the telencephalon and diencephalon. Another landmark at this stage of development is the depth of the falx cerebri (fc and asterisks in Fig. 36), which partitions the two halves of the septum and distinguishes the septal region derived from the paired lateral ventricles and the preoptic area derived from the midline third ventricle.

The pattern of neurogenesis in the rostral portion of the preoptic area is illustrated in radiograms from rats labeled with [^{3}H]thymidine between days E13 and E16 and killed on day E20 (Figs. 37, 38). Three major components may be distinguished, the anterolateral preoptic region (PAL), the superior preoptic region (PS), and the anteromedial preoptic region (PAM). The *anterolateral preoptic region* is the earliest-produced component. Following injection on day E13 (Figs. 37A, 39) it is composed of some unlabeled and many more

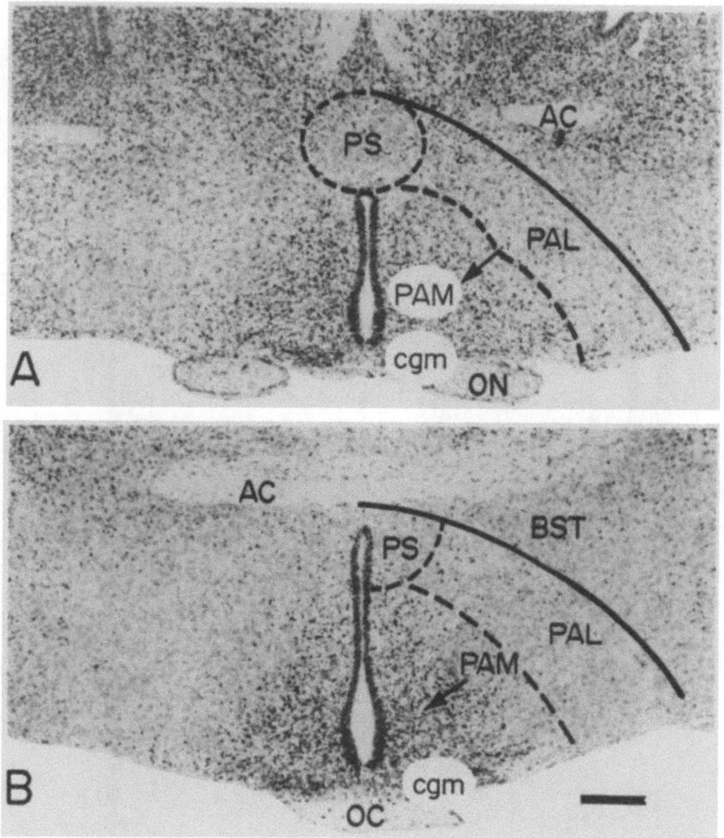

Fig. 38. Legend see Fig. 37

heavily labeled neurons. Following injection on day E14 (Figs. 37B, 40) most of the cells are unlabeled; some are heavily labeled, with an apparent neurogenetic gradient from dorsal to ventral (arrow in Fig. 40). This suggests that the cells of the anterolateral preoptic region may originate from the lateral portion of the rostral wall of the optic fissure and settle in an outside-in sequence. (The lateral portion of the optic fissure recedes after the penetration of the optic tract fibers). Accordingly, peak production time of the neurons of the anterolateral preoptic region is on day E13. The *superior* (*median*) *preoptic region* is situated medially beneath the anterior commissure. It is composed of lightly labeled cells in the rat injected on day E13 (Figs. 37A, 39), of many heavily labeled cells in the rat injected on day E14 (Figs. 37B, 40), and mostly of unlabeled cells in the rat injected on day E15 (Fig. 38A). Therefore, the peak production time of neurons of the superior preoptic region is on day E14. The cell-dense and large *anteromedial preoptic region* is composed of neurons that are lightly labeled after injection on day E13 (Figs. 37A, 39). It contains heavily labeled cells, preferentially in a peripheral position, after injection on day E14 (Figs. 37B, 40). After injection on day E15 most of the anteromedial preoptic neurons are heavily labeled (Fig. 38A). The area occupied by labeled cells progressively shrinks medially after injection on day E16 (Fig. 38B) and

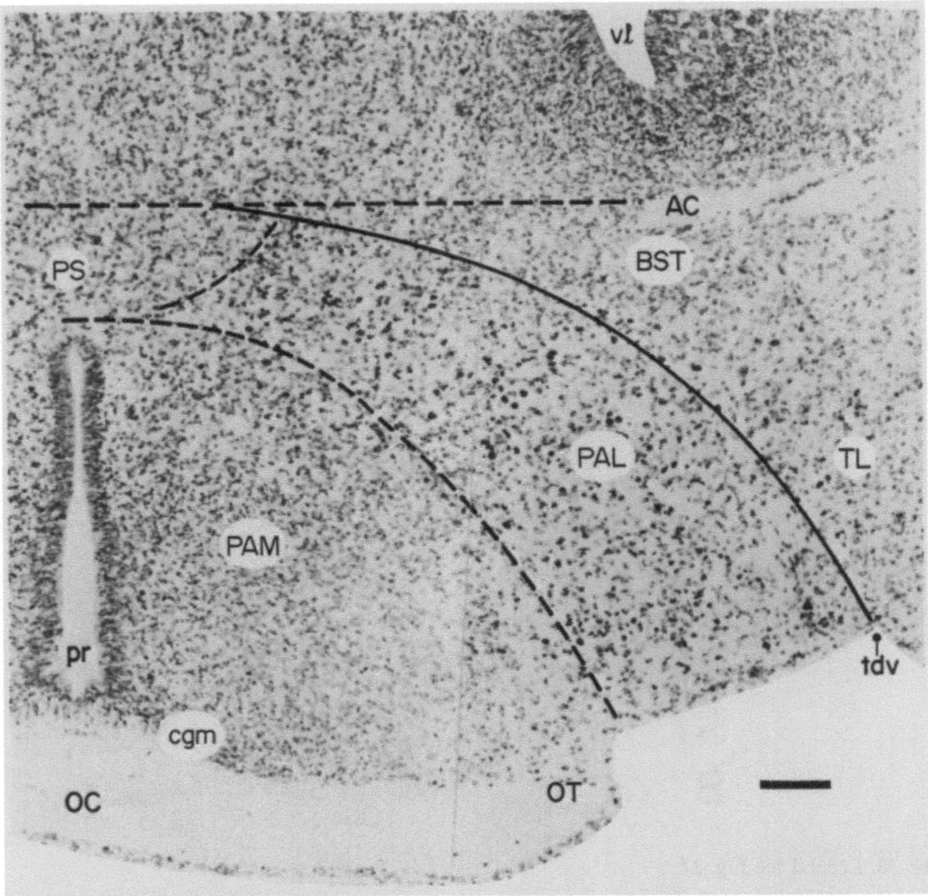

Figs. 39 and 40. Coronal thymidine radiograms at higher magnification of the rostral portion of the preoptic area in perinatal rats injected on day E13 (Fig. 39) and E14 (Fig. 40). Note the changing labeling pattern in the anterolateral, superior, and anteromedial preoptic regions. Paraffin. *Scale*, 400 μm

day E17 (Fig. 41 A). In perinatal rats injected on day E18 the labeled cells remain near the ventricle (not shown). Accordingly, the neurons of the anteromedial preoptic region are generated between days E15 and E17 and settle in an outside-in sequence from lateral to medial.

The structure of the preoptic area is more complex posteriorly than anteriorly, as seen in coronal radiograms from a rat injected on day E17 and killed on day E20 (Figs. 41, 42). The simple concentric arrangement of heavily labeled cells at anterior levels (Fig. 41 A) changes progressively (Fig. 41 B) into a butterfly pattern in the posterior preoptic area (Fig. 42A). This pattern persists into the transition region of the suprachiasmatic region of the anterior hypothalamus (Fig. 42B). Because the superior (median) preoptic region does not change from rostral to caudal levels, it is designated by the same name throughout the preoptic area. But the lateral region becomes smaller and changes its position posteriorly and is, therefore, tentatively designated as a separate region, the *posterolateral preoptic region* (PPL). The posterior continuation of the anterome-

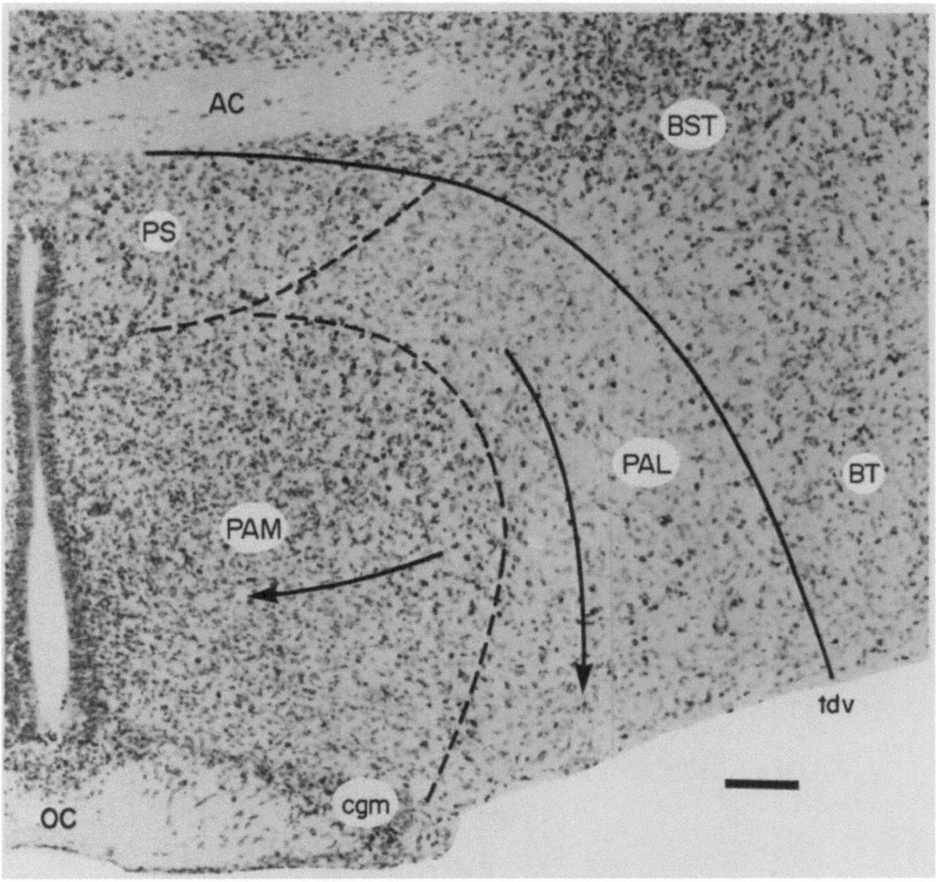

Fig. 40. Legend see Fig. 39

dial preoptic area forms two wings or horns that are reminiscent of the dorsomedial and ventromedial nuclei of the hypothalamus (Sect. 6.3). They are designated as the *posterodorsal* (PPD) and *posteroventral* (PPV) *preoptic regions* (Figs. 41 B, 42 A, 43). Radiograms from rats killed on day E19 revealed that injection on day E15 (Fig. 44 A) labels lightly or heavily most of the neurons of the posterodorsal and posteroventral preoptic regions; injection on day E16 (Fig. 44 B) results in heavy labeling of most cells except those situated peripherally in the two horns; injection on day E17 (Figs. 41 B, 42 A, 44 C) limits labeling to the base of the horns, with some variability in different planes and in different animals. After injection on day E18 (not shown) the few cells that are labeled are confined to the vicinity of the ventricle. From these observations it is concluded that the neurons of the posterodorsal and posteroventral preoptic regions are generated between day E15 and E17 with a dorsolateral-to-ventromedial gradient in the posterodorsal region and a less-pronounced mirror image gradient in the posteroventral region.

The area between the two horns of the posterodorsal and posteroventral preoptic regions is occupied by a spherical aggregate of earlier-generated neu-

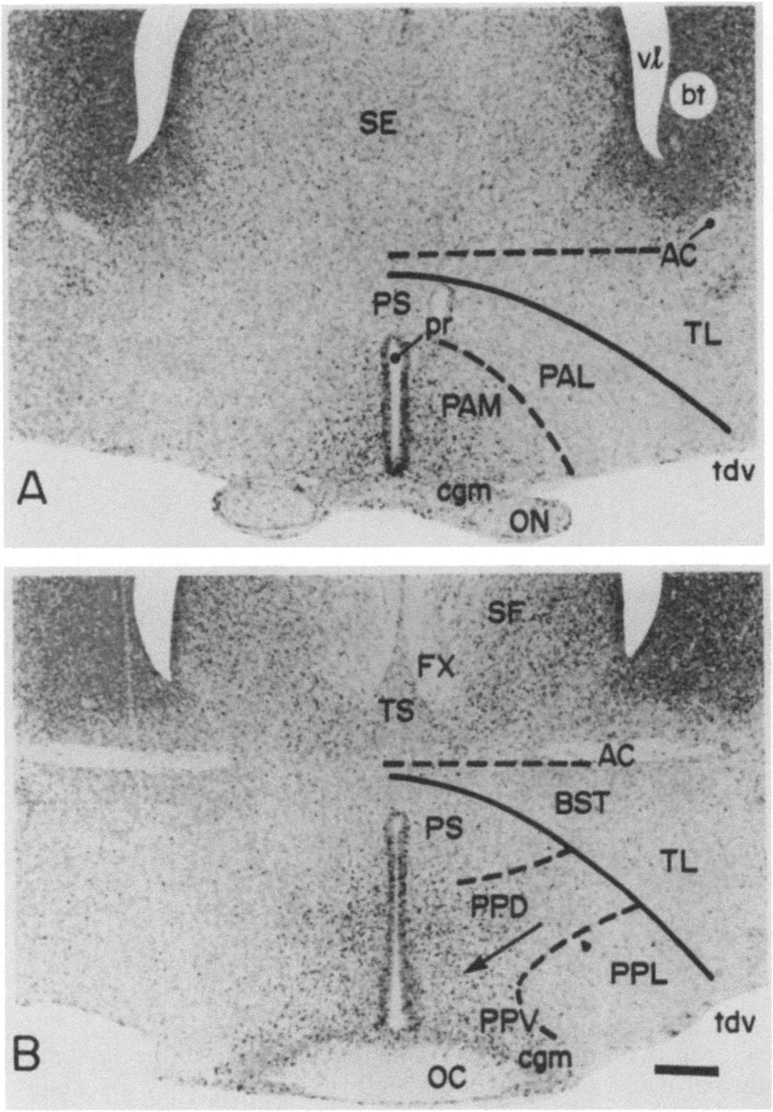

Figs. 41 and 42. Coronal thymidine radiograms of the preoptic area, from rostral (Fig. 41 A) to caudal (Fig. 42 B), in a perinatal rat labeled on day E17. Note the difference in cytological structure and labeling pattern in the rostral portion of the preoptic area (Fig. 41 A), its caudal portions (Figs. 41 B, 42 A), and the transition area to the anterior hypothalamus (Fig. 42 B). Paraffin. *Scales*, 200 μm

rons, designated as the *intermediate preoptic region* (PIM). Many of these cells are still labeled with injections made on day E15 (Fig. 44 A) but very few after injection on day E16 (Figs. 43, 44 B). The production time of the neurons of the intermediate preoptic region is intermediate between the early-generated neurons of the posterolateral preoptic area and the late-generated neurons of the posterodorsal and posteroventral preoptic areas.

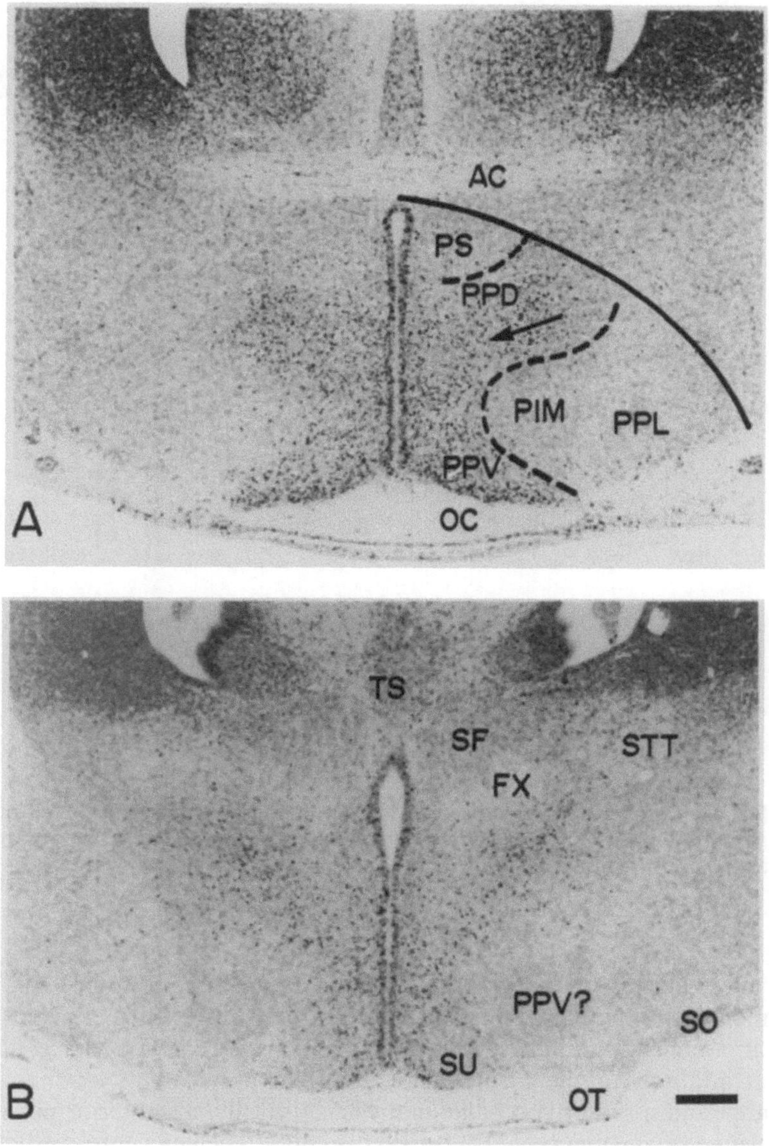

Fig. 42. Legend see Fig. 41

Comments. The neurons of the preoptic area, as a whole, are generated between days E12 and E17 and settle in a modified outside-in sequence. Both the chronology and the settling pattern agree with our earlier quantitative results in adult rats with long-survival thymidine radiography (Altman and Bayer 1978a: Fig. 5). However, since in our earlier study we used a simple subdivision of the preoptic area we cannot correlate the details of the present results in late embryos with the final adult pattern. The preoptic area was often treated in the past as a single homogeneous region (e.g., de Groot 1959) or one divisible into two parts, a lateral and a medial part (König and Klippel 1963). Because of the functional

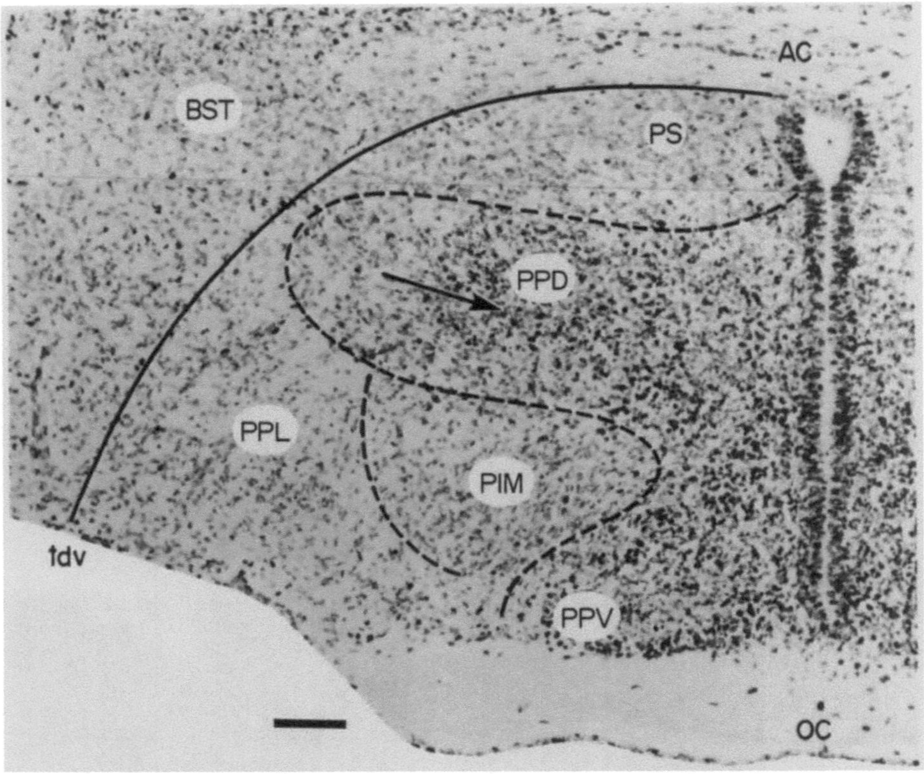

Fig. 43. Coronal thymidine radiogram from a rat injected on day E16 and killed on day E19 to show the complex configuration of components of the posterior portion of the preoptic area. Paraffin. *Scale*, 400 μm

heterogeneity of the preoptic area, and evidence of some regional sexual dimorphism, there have been subsequent attempts to further subdivide this structure. One of the most detailed subdivisions is that of Bleier et al. (1979), who, in addition to the medial and lateral preoptic areas, distinguished a dorsal area and a septohypothalamic nucleus dorsally, an anterior periventricular nucleus dorsomedially, and a medial preoptic nucleus ventromedially. Bleier and her associates considered the preoptic area to be merely a continuation of the anterior hypothalamus, not a separate region. Indeed, the two divisions that we have designated as the posterodorsal and the posteroventral preoptic regions (Figs. 41 B, 42 A, 43) are referred to as the anterior hypothalamic nucleus and the anterior hypothalamic area, respectively (Bleier et al. 1979: Plates 7–10).

Our present observations indicate that neurogenetically the preoptic area is a discrete division of the diencephalon that can be demarcated caudally from the anterior hypothalamus, and rostrally, dorsally, and laterally from neighboring telencephalic structures. The preoptic area can be defined as a portion of the hypothalamus derived from the neuroepithelium of the rostral wall of the optic fissure of the third ventricle. This neuroepithelial region extends in the basal horizontal plane from the optic fissure ventrolaterally to the tip of the preoptic recess anteromedially, and in the dorsal horizontal plane from the

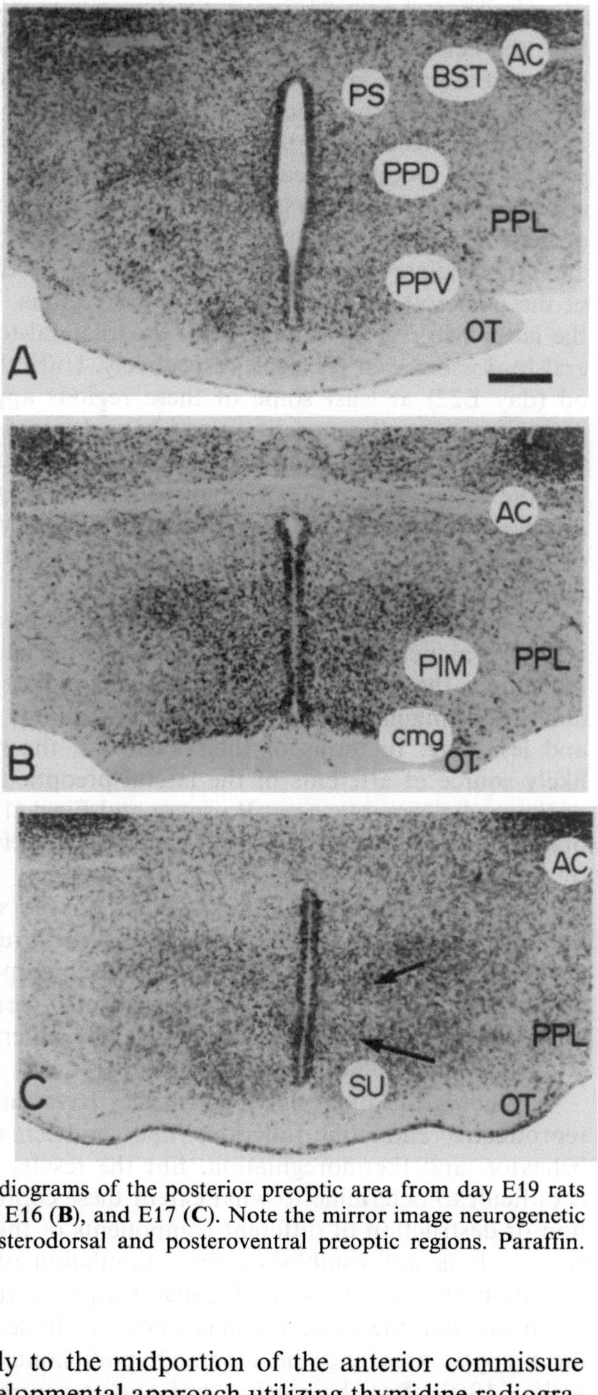

Fig. 44 A–C. Coronal thymidine radiograms of the posterior preoptic area from day E19 rats that were labeled on days E15 (**A**), E16 (**B**), and E17 (**C**). Note the mirror image neurogenetic gradients (*arrows in* **C**) in the posterodorsal and posteroventral preoptic regions. Paraffin. *Scale*, 200 μm

foramen of Monro posteriorly to the midportion of the anterior commissure anteriorly. Moreover, the developmental approach utilizing thymidine radiography allows a subdivision of the preoptic area into many more regions than was hitherto possible. The preoptic area can be divided into an anterior and posterior part on the basis of structural complexity during early development. The anterior part of the preoptic area comprises three neurogenetically distinct regions, the anterolateral, the superior (median), and the anteromedial. The neurons of the anteromedial preoptic region are produced with a peak on day E13, the neurons of the superior (median) preoptic region with a peak on day E14, and the neurons of the anteromedial preoptic region mostly between days E15 and E17. In the latter region there was a clear lateral-to-medial gradient, that is, the cells disperse and settle in a radial outside-in pattern. The posterior part of the preoptic area has five components, the posterolateral, superior, intermediate, posterodorsal, and posteroventral regions. The neurons of the posterolateral region are produced first, followed by the neurons of the superior and intermediate regions, and ending with the production and settling of neurons of the posterodorsal and posteroventral regions. The neurogenetic gradient in the latter two regions is an oblique dorsolateral-to-ventromedial and ventrolateral-to-dorsomedial gradient, respectively. Until the end of the embryonic period (day E22) at least some of these regions appear to be "nuclei," that is, are uniform in cell composition and have delineable boundaries. However, they are designated as "regions" because as development proceeds the cells disperse and the boundaries of the originally discrete cell aggregates are no longer recognizable. But it is expected that because of the different times of origin of neurons of the different regions their boundaries should be definable in adults with long-survival thymidine radiography.

The evidence that the relatively homogeneous-appearing adult preoptic area is neurogenetically quite heterogeneous is not surprising in view of available indications that different parts of this area may receive different afferent inputs and are distinguishable by cytochemical methods, physiological manipulations, and lesioning. In terms of their locations, the medial forebrain bundle is a likely source of afferents of the lateral preoptic area, and the stria terminalis of the medial preoptic area. Raisman and Field (1973) described in the anterior preoptic area of the rat a "strial part" that receives fibers from the amygdala by way of the stria terminalis, and an adjacent "round nucleus" which does not. The former appears to be located in the superior preoptic region and the latter in the dorsal part of the anterolateral region. In a recent study, Ruggiero et al. (1984) showed that in the mouse preoptic area the tyrosine hydroxylase-containing cells (catecholamine neurons) clearly outline the superior preoptic region or what they refer to as the anterior preoptic nucleus of Loo (Ruggiero et al. 1984: Figs. 2–4).

Physiological and behavioral studies have implicated the preoptic area in reproductive endocrine functions, male and female sexual behavior, maternal behavior, and thermoregulation. But the results are often inconsistent and it is difficult to determine to what extent these contradictions are due to stimulation or destruction of different components of the preoptic area in the various studies. It is well established that stimulation of the preoptic area produces ovulation (Everett 1964) and gonadotropin hormone release (Clemens et al. 1976), and that these effects can be abolished by severing the connection between the preoptic area and the tuberal hypothalamus (Tejasen and Everett 1967; Halász 1969). But the mechanisms and structures involved are controversial. Some immunocytochemical studies have localized the gonadotropin-releasing factor in the preoptic area (e.g., Barry et al. 1974; Witkin et al. 1982), but others have questioned its presence in the cell body of neurons (Baker et al. 1975) or the preoptic origin of the axons containing this releasing hormone (Merchenthaler et al. 1980). Several studies have reported that large preoptic lesions, particularly if they involve the suprachiasmatic nucleus, abolish the estrus cycle and ovulation (Halász 1969; Barraclough 1973). But the effects of smaller lesions are controversial. Clemens et al. (1976) reported that preoptic lesions that did not extend into the anterior hypothalamus did not interfere with ovulation, though it disturbed the estrus cycle by producing pseudopregnancies. The implication is that ovulation is not controlled by the preoptic area. But this was not confirmed by Gray et al. (1978). Wiegand et al. (1980) reported that small lesions apparently restricted to what we are describing as the anteromedial preoptic area produced permanent estrus but this did not result from lesions in any other region of the preoptic area. A study by Shander and Barraclough (1980) also implicated this area, what they described as the medial basal anterior preoptic region, in the regulation of the estrus cycle and the preovulatory surges of luteinizing hormone and follicle-stimulating hormone.

The role of the preoptic area in the regulation of sexual behavior is also inconsistent. In the male rat preoptic lesions interfere with sexual behavior (Heimer and Larsson 1966/1967; Lisk 1968) and stimulation of the preoptic area facilitates sexual responses (Vaughn and Fisher 1962; Malsbury 1971). Implantation of testosterone (Lisk 1967) or even estrogen (Christensen and Clemens 1974) promotes male sexual activity, and implantation of estrogen into the preoptic area of the female induces female sexual behavior (Lisk 1962). However, unlike in the male, destruction of the preoptic area in the female does not interfere with sexual behavior (Law and Meagher 1958; Singer 1968; Powers and Valenstein 1972) even though this region has been shown to have a high affinity for the uptake of estrogen (Stumpf 1968; Pfaff and Keiner 1973; Stumpf and Sar 1975). Although preoptic lesions do not affect the sexual receptivity of the female, maternal behavior is severely interrupted even if the small lesion is restricted to, what appears to be, the anteromedial preoptic region (Numan 1974: Fig. 4).

The issue of a better and more consistent parcellation of the preoptic area is particularly relevant regarding the localization of its so-called sexually dimorphic components. In the study referred to earlier, Raisman and Field (1973) found that in the "strial part" of the preoptic area the number of synapses of nonamygdaloid origin was higher in female rats than in males. Later Gorski et al. (1978) reported that the volume of the medial preoptic nucleus (what appears to be our posterodorsal preoptic region) is as much as eight times larger in the male rat than in the female. The differential increase in volume became evident during the perinatal period (Jacobson et al. 1980) and was, therefore, presumably due to some factor other than differential cell production. In a more recent comparative study (Bleier et al. 1982) a much more extensive sexually dimorphic region is claimed to exist in several species, including the rat, on the basis of qualitative observations.

6.2 Development of the Anterior Hypothalamus: Rostral Region

The anterior hypothalamus is derived from that portion of the hypothalamic neuroepithelium which, in the midline in young embryos, is delineated by the *optic fissure* (*of*) rostrally and the infundibular recess (ir) caudally (Figs. 3,

7). Away from the midline, the anterior hypothalamic rudiment extends dorsally (ahd in Figs. 3B, 7B, 8A), and laterally it becomes contiguous with the rudiment of the anterior thalamus (ta in Figs. 4B, 8B). For convenience we have divided the anterior hypothalamic rudiments and its derivatives into a rostral and caudal component. The rostral portion of the anterior hypothalamic rudiment (ahr in Figs. 3A, 4A, B, 8A, B) faces medially the optic fissure and forms its caudal wall. The caudal portion of the anterior hypothalamic rudiment (ahr) faces medially the infundibular recess and is laterally contiguous with the ventral portion of the posterior hypothalamus (phv in Figs. 3B, 4A, B, 8B). The major derivatives of the rostral region of the anterior hypothalamic neuroepithelium to be dealt with in this section are the components of the anterior hypothalamic nuclei, the supraoptic nucleus, and the paraventricular nucleus.

6.2.1 The Anterior Hypothalamic Nuclei

Background. The bulk of the anterior hypothalamus, apart from the clearly defined suprachiasmatic, paraventricular, and supraoptic nuclei, has a diffuse appearance in adult rats, and is usually referred to as a single entity under the name of anterior hypothalamic area or anterior hypothalamic nucleus (Gurdjian 1927: Figs. 3, 4; Krieg 1932: Fig. 1; de Groot 1959: Plate 2; König and Klippel 1963: Figs. 24–30; Christ 1969: Fig. 15). Recently, Bleier et al. (1979) have attempted to subdivide this region and proposed the following components: the anterior area, the anterior nucleus, the dorsal area, the lateral area, the caudal anterior area, the dorsal tuberal nucleus, and the periventricular area. In an earlier study of the embryonic development of the hypothalamus we have tentatively identified an early-generated and early-differentiating region, the anterobasal nucleus (Altman and Bayer 1978b: Figs. 8, 10). The present study has confirmed the neurogenetic distinctiveness of this nucleus and allowed us to identify several other anterior hypothalamic nuclei.

Results. In rats injected with [^{3}H]thymidine on day E13 and killed on day E14, an aggregate of unlabeled cells is found beneath the midline neuroepithelium (AB in Fig. 45) at the level of the optic fissure and caudal to it. This complement of cells, which must have been produced before the morning of day E13, is associated with a fibrous zone, possibly composed of fibers of the supraoptic decessation (SD?). This region we designate as the *anterobasal nucleus*. In rats injected on day E13 and killed on day E20, the anterobasal nucleus situated caudal to the optic chiasm contains a high proportion of labeled neurons (Fig. 46A). Evidently, the migration of cells forming after the morning of day E13 continues into the anterobasal nucleus from its neuroepithelium (ab in Fig. 45). In rats injected on day E14 and killed on day E15 (Fig. 47) the bulk of the anterobasal nucleus is composed of unlabeled cells but dorsolaterally (arrow) the migration of labeled cells is still indicated. In rats injected on day E14 and killed on day E22 (Figs. 48, 49) the labeled cells appear to be mingled with unlabeled cells, indicating an absence of neurogenetic gradient in the anterobasal nucleus. In rats injected on day E15 and killed on day E20 (Fig. 50) or E22 (Fig. 51) there are few labeled cells in the anterobasal nucleus. This establishes that the neurons of the anterobasal nucleus are produced predominantly on days E13 and E14.

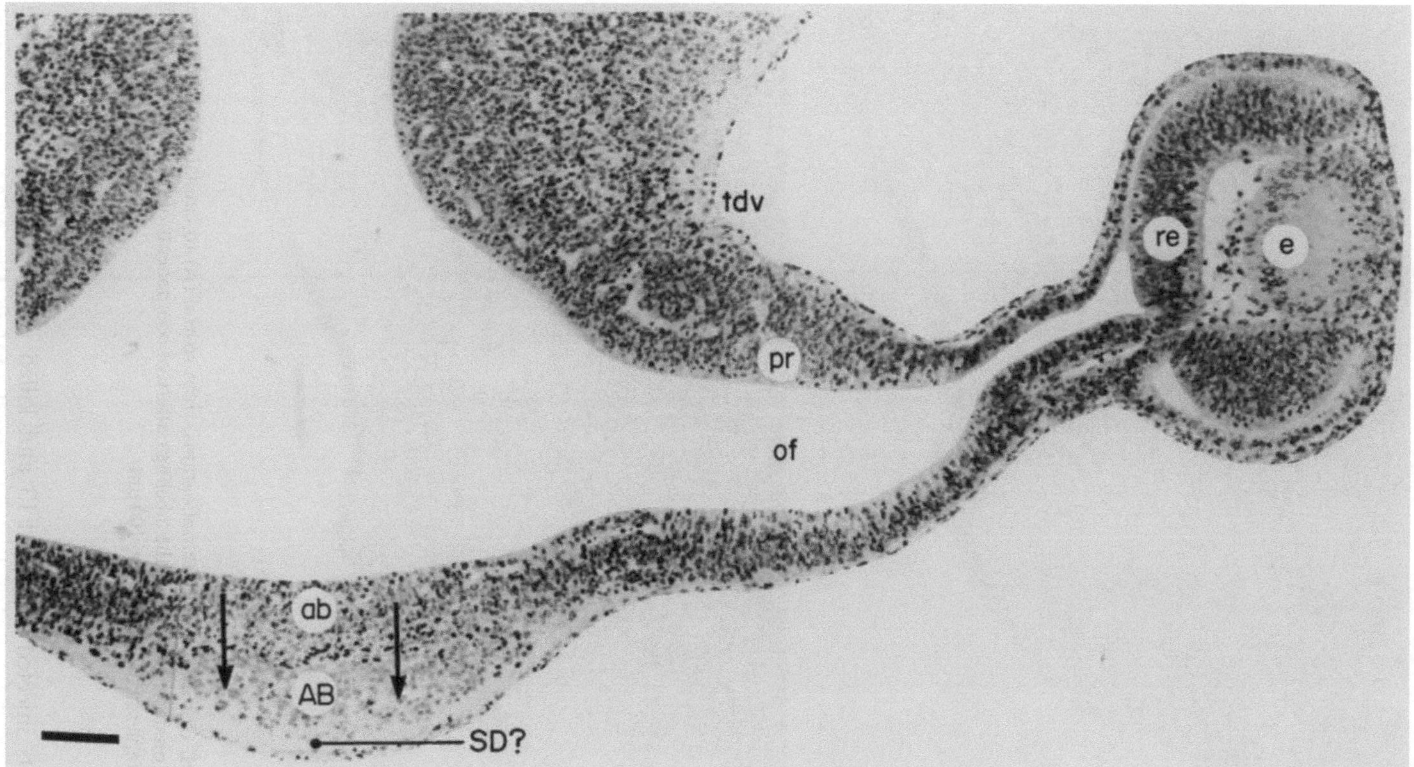

Fig. 45. Coronal thymidine radiogram from a rat labeled on day E13 and killed on day E14 to show the early differentiating neurons of the anterobasal nucleus. *Arrows* indicate the presumed migratory route from the neuroepithelial zone of the nucleus. Methacrylate. *Scale*, 100 μm

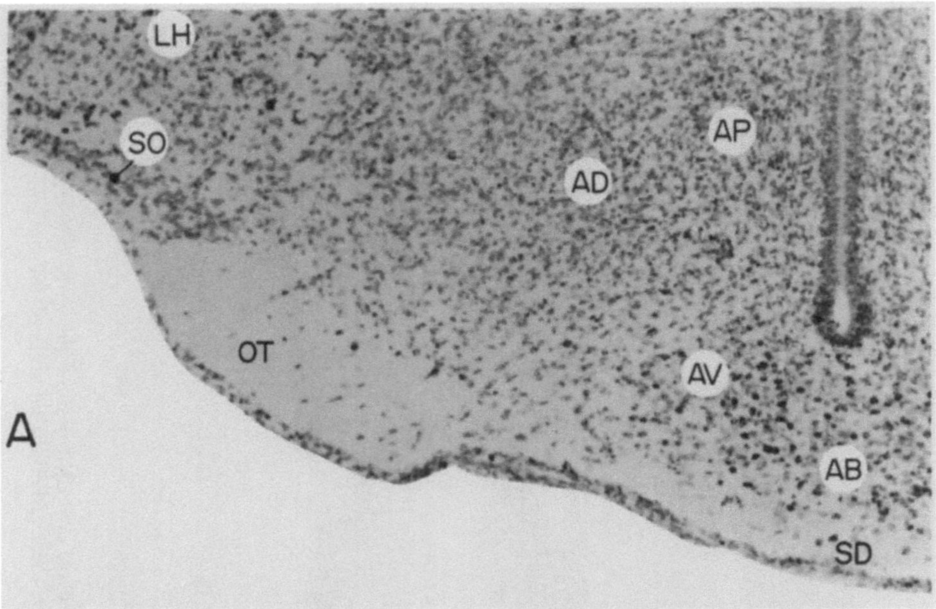

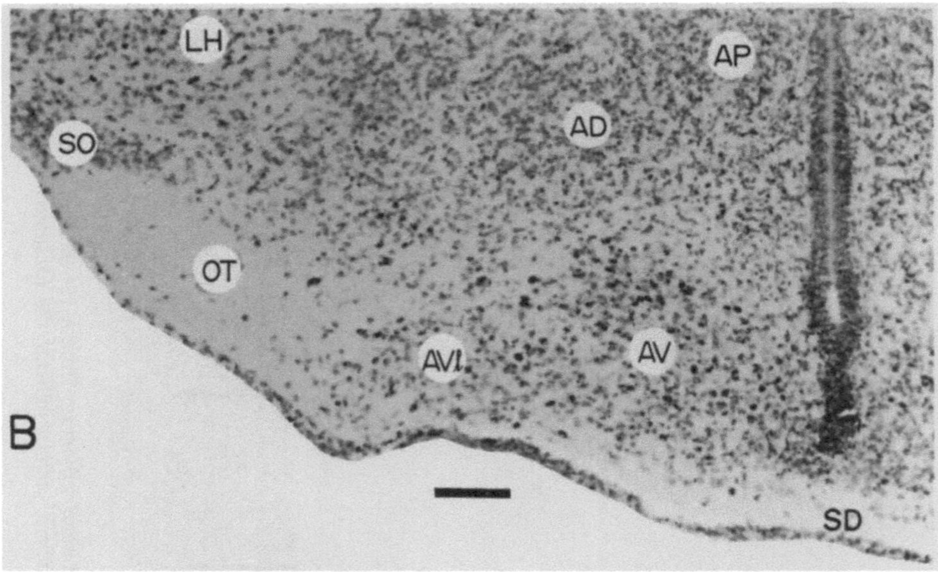

Fig. 46 A, B. Coronal thymidine radiograms, from rostral (**A**) to caudal (**B**), from a perinatal rat labeled on day E13 to show the labeling pattern of components of the anterior hypothalamic nuclear complex. Paraffin. *Scale*, 100 μm

In rats injected on day E13 and killed on day E20 heavily labeled cells predominate in a region located lateral (AV in Fig. 46A) and posterolateral (AV in Fig. 46B) to the anterobasal nucleus. Although this region could be considered the lateral extension of the anterobasal nucleus, we designate it as a separate structure, the *anteroventral nucleus,* for the following reasons. First, unlike the anterobasal nucleus, which contains either heavily labeled cells or

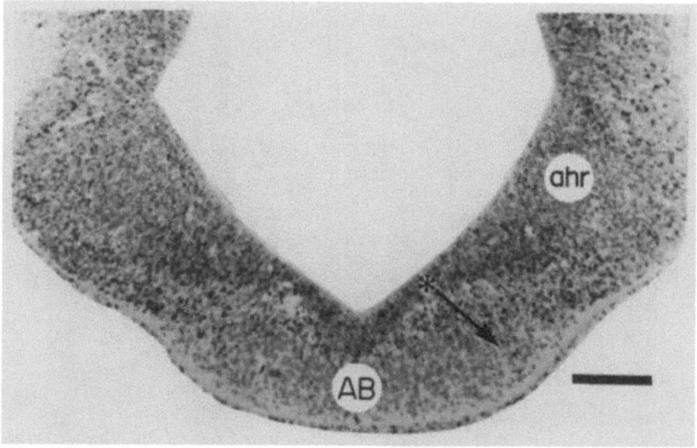

Fig. 47. Coronal thymidine radiogram from a rat injected on day E14 and killed on day E15 showing the continuing migration of labeled cells dorsolateral to the anterobasal nucleus (*arrow*). Methacrylate. *Scale*, 200 μm

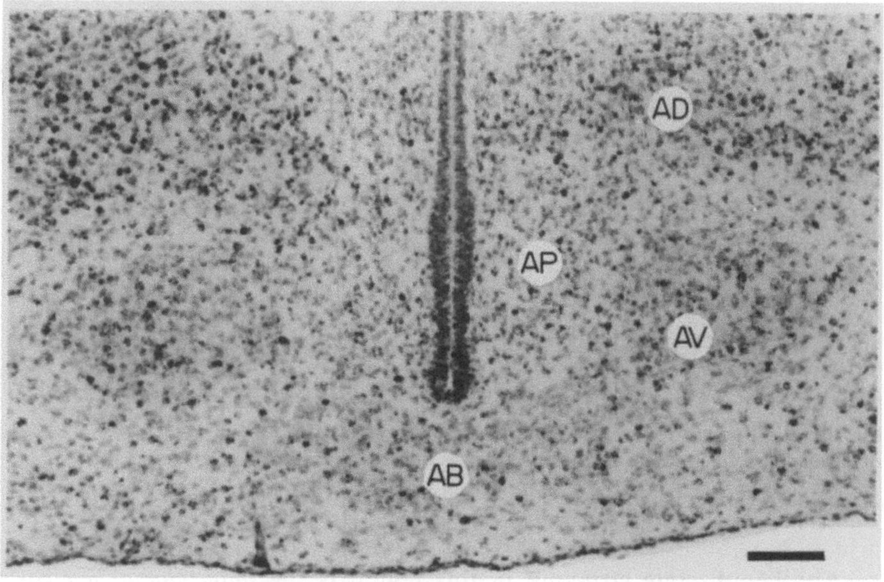

Fig. 48. Labeling pattern in the anterobasal, anteroventral, and anterodorsal nuclei in a perinatal rat tagged on day E14. Paraffin. *Scale*, 100 μm

unlabeled cells in rats injected on day E14 and killed on day E22 (AB in Fig. 48), the anteroventral nucleus is mainly composed of heavily or lightly labeled cells (AV in Figs. 48, 50). Second, while the anterobasal nucleus has no neurogenetic gradient, there seems to be a dorsal-to-ventral gradient in the anteroventral nucleus in rats injected on day E15 and killed either on day E20 (arrow in Fig. 50) or day E22 (arrow in Fig. 51). Finally, the cytological structure of the anteroventral nucleus appears to be different than that of the anterobasal nucleus (Fig. 51).

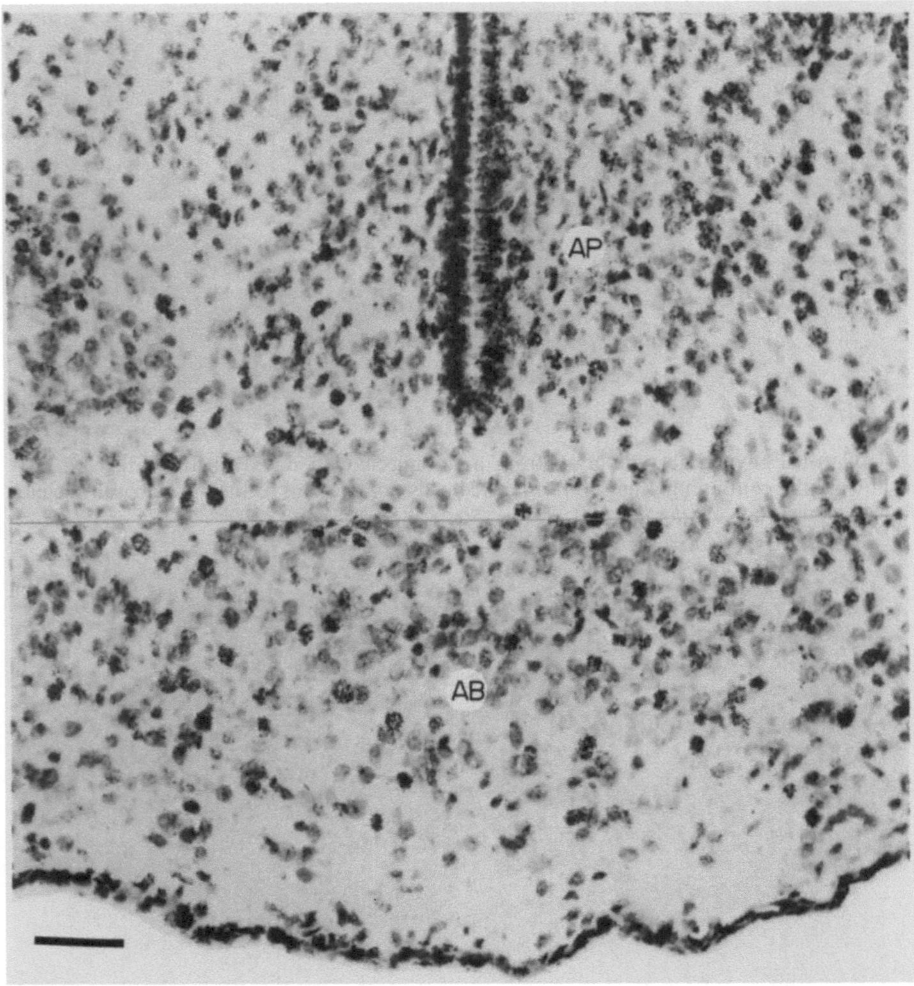

Fig. 49. High-power radiogram from a perinatal rat labeled on day E14 to show the absence of a neurogenetic gradient in the anterobasal nucleus. Paraffin. *Scale*, 50 μm

Another discrete structure in this region is the *anterodorsal nucleus* (AD in Figs. 46, 48, 50, 51). The anterodorsal nucleus is situated ventral to the paraventricular nucleus, lateral to the anterior periventricular area (AP), and dorsolateral to the anteroventral nucleus. The cells of the anterodorsal nucleus are lightly labeled in perinatal rats that received [^{3}H]thymidine on day E13 (Fig. 46) and are heavily labeled in rats injected on day E14 (Fig. 48); and it contains a higher proportion of heavily labeled cells than the anteroventral nucleus in rats injected on day E15 (Figs. 50, 51). Accordingly, the neurons of the anterodorsal nucleus are generated predominantly on days E14 and E15 with an indication of a ventral-to-dorsal neurogenetic gradient (Figs. 50, 51). The most likely source of the anteroventral and anterodorsal nuclei is the neuroepithelium lateral to that of the anterobasal nucleus neuroepithelium (asterisk in Fig. 47). A less-distinct anterior hypothalamic nucleus is situated medial to

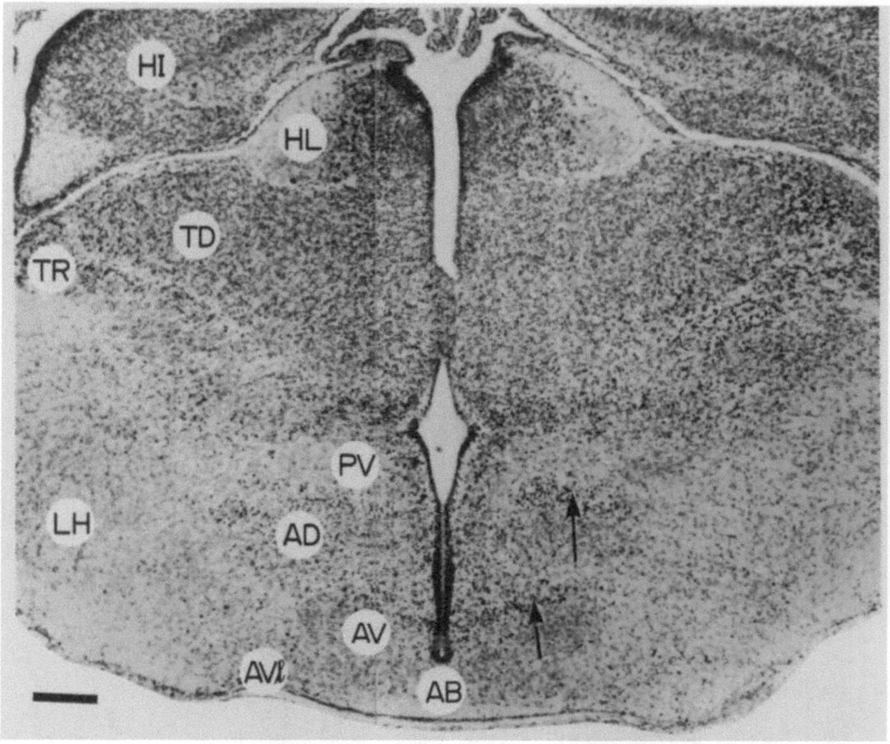

Fig. 50. Coronal thymidine radiogram from a perinatal rat labeled on day E15 to show the ventral-to-dorsal neurogenetic gradients (*arrows*) in the anteroventral and anterodorsal nuclei. Paraffin. *Scale*, 200 μm

the optic tract and is designated as the *lateral anteroventral nucleus* (AVl in Figs. 46B, 50, 51).

Comments. The anterobasal nucleus situated immediately caudal to the optic chiasma is not usually recognized as a discrete structure. It is either unnamed (de Groot 1959: Plate 2, level A6.8) or is designated as part of the arcuate nucleus (Christ 1969: Fig. 15). Bleier et al. (1979: Plate 15) refer to it as the infundibular nucleus, a term that they use as a synonym of the arcuate nucleus. We have identified the anterobasal nucleus in our previous study of hypothalamic development (Altman and Bayer 1978b: Figs. 8–10) but did not clearly distinguish it from the more caudally situated arcuate nucleus in our quantitative study of neurogenesis (Altman and Bayer 1978a: Fig. 6). Two lines of evidence establish the distinctiveness of the anterobasal nucleus. First, the arcuate nucleus is composed in adults of small round, densely packed cells, while the anterobasal nucleus is composed of somewhat larger, spindle-shaped, loosely packed cells. Second, the neurons of the arcuate nucleus are among the latest-produced cells of the hypothalamus (Sect. 7.2), whereas the neurons of the anterobasal nucleus are among its earliest-produced cells.

In the present study we have obtained neurogenetic evidence for the existence of at least two additional cell aggregates in this region: the anteroventral and

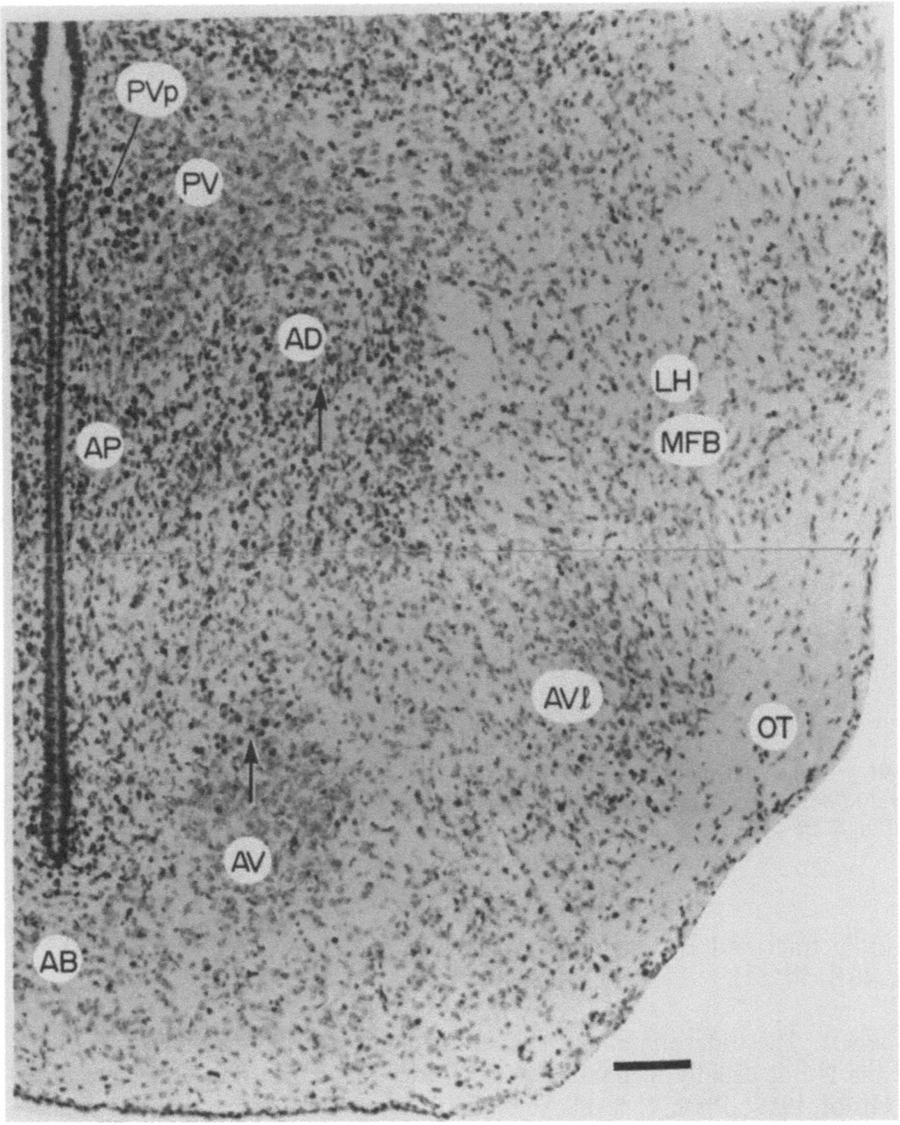

Fig. 51. High-power thymidine radiogram from a perinatal rat labeled on day E15 to show the different labeling patterns and neurogenetic gradients (*arrows*) of the components of the anterior hypothalamic nuclear complex. Paraffin. *Scale*, 100 μm

the anterodorsal nuclei. These nuclei form separate clusters and are composed of neurons that are generated later than those of the anterobasal nucleus. The internuclear neurogenetic gradient in this region appears to be: anterobasal nucleus → anteroventral nucleus → anterodorsal nucleus. In the anteroventral and anterodorsal nuclei, but not in the anterobasal nucleus, there are indications also for a ventral-to-dorsal intranuclear neurogenetic gradient (arrows in Figs. 50, 51). The later-produced neurons of the anteroventral and anterodorsal nuclei are apparently derived from a germinal site that is situated dorsolateral (Fig. 47) to the neuroepithelium of the anterobasal nucleus (Fig. 45).

In relation to the detailed subdivision of the adult anterior hypothalamus by Bleier et al. (1979: Plates 11–15), which does not distinguish sharply between this region and the preoptic area, it seems that the anteroventral nucleus corresponds to their caudal anterior nucleus, the anterodorsal nucleus corresponds to what they designate as the anterior area, and our lateral anteroventral nucleus corresponds to their lateral anterior nucleus.

6.2.2 The Paraventricular Nucleus

Background. The neurons of the paraventricular nucleus synthesize the nonapeptides, oxytocin, and vasopressin, in association with an acidic protein, neurophysin, and convey these substances to the posterior (neural) lobe of the pituitary gland. The paraventricular nucleus is divided into two major parts, a conspicuous magnocellular division and a parvocellular division (Gurdjian 1927; Krieg 1932). Hatton et al. (1976) and Armstrong et al. (1980) further subdivided the magnocellular paraventricular nucleus into medial, lateral, and posterior parts. The subdivisions referred to by Swanson and Sawchenko (1983) are the anterior, medial, and posterior parts. There is immunohistochemical evidence that the oxytocin-staining (synthesizing) and vasopressin-staining cells are segregated in the magnocellular paraventricular nucleus (Swaab et al. 1975; Choy and Watkins 1979; Dierickx 1980; Rhodes et al. 1981; Sawchenko and Swanson 1983). Armstrong et al. (1980) suggested that there is a higher proportion of oxytocin cells in the medial magnocellular subnucleus, and of vasopressin cells in the lateral magnocellular subnucleus. Rhodes et al. (1981) interpreted their results to indicate that the medial subnucleus contained primarily oxytocin-producing cells, while the lateral subnucleus consisted of a core of vasopressin-producing cells with a few oxytocin cells forming a rim around the vasopressin cells. Swanson and Sawchenko (1983) suggested that the anterior and medial magnocellular parts are composed almost exclusively of oxytocin-staining cells and in the posterior magnocellular part the oxytocin cells are concentrated anteroventromedially and the vasopressin cells posterodorsolaterally. They also noted that the vasopressin-staining cells tend to be larger than the oxytocin-staining cells.

The parvocellular paraventricular nucleus was subdivided by Armstrong et al. (1980) into a dorsomedial cap and a posterior subnucleus. Of the five subdivisions of the parvocellular nucleus proposed by Swanson and Sawchenko (1983) at least the periventricular part needs to be added to the two distinguished by Armstrong et al. The neurons of the parvocellular nucleus do not contain neurophysin but many of them synthesize somatostatin (Dierickx 1980). Together with the large neurons of the posterior nucleus, the parvocellular elements located in this region may be the source of the paraventricular projection to the brain stem and the spinal cord (Hosoya and Matsushita 1979; Armstrong et al. 1980; Swanson and Sawchenko 1983).

Using the technique of a single injection of [^{3}H]thymidine, Ifft (1972) found the highest concentration of heavily labeled magnocellular paraventricular neurons on day E14, and relatively few on day E15. Anderson (1978), using the same technique, found peak production of magnocellular neurons on day E14 and a high percentage of heavily labeled cells still present following injection on day E15. In our study with the double-labeling procedure, we reported

(Altman and Bayer 1978a: Fig. 6) that slightly more magnocellular neurons are produced on day E15 than on day E14. In none of these studies was an attempt made to study differences in the time of origin of magnocellular subnuclei but we noted a dorsolateral-to-ventromedial gradient. All of these studies agree that the smaller neurons of the parvocellular region are produced after the large magnocellular neurons.

In recent years several studies dealt with the developmental aspects of oxytocin, vasopressin, and neurophysin synthesis in the brain. Sinding et al. (1980), using a radioimmunoessay procedure, detected neurophysin as early as day E13 in whole blocks of rat hypothalamic and pituitary tissue. They found oxytocin first on day E14. Choy and Watkins (1979), using immunocytochemical procedures, have first detected neurophysin in the rat hypothalamus on day E18, vasopressin on day E19, and oxytocin not until postnatal day 1–2. Khachaturian and Sladek (1980) found neurophysin-staining in the paraventricular nucleus by day E17. Finally, Whitnall et al. (1985) reported that both oxytocin-associated neurophysin and vasopressin-associated neurophysin could be detected in the paraventricular nucleus by day E17. The latter authors provided maps of the distribution of oxytocin-associated and vasopressin-associated neurophysin-staining neurons in the developing hypothalamus of the rat.

Results. The cellular differentiation of the paraventricular nucleus is not sufficiently advanced in perinatal rats unequivocally to distinguish its magnocellular and parvocellular components. Accordingly, we begin our analysis with the labeling pattern seen in adult (P60) rats that received [^{3}H]thymidine on days E15+E16 (Figs. 52–55). We shall designate the region referred to as lateral by Armstrong et al. (1980) and as anterior by Swanson and Sawchenko (1983) as the *anterolateral subnucleus* (PVa). Most rostrally (Fig. 52A) the anterolateral nucleus is composed predominantly of unlabeled cells. At this level few magnocellular neurons are seen in the rest of the paraventricular nucleus. In the next section caudally (Fig. 52B), the mostly unlabeled neurons of the anterolateral subnucleus form a crescent-shaped envelope around the *medial subnucleus* (PVm). Neurons in the medial subnucleus and those in the *posterior subnucleus* (PVc) tend to be labeled. Still more caudally (Fig. 53A), the few neurons that may belong to the anterolateral subnucleus are mostly unlabeled while the neurons of the other subnuclei tend to be labeled. In the most caudal section (Fig. 53B), the neurons of the posterior subnucleus are typically labeled. Throughout the paraventricular nucleus the great majority of the neurons of the *periventricular subnucleus* (PVp) are labeled in this injection group. An examination of the unlabeled and labeled cells at higher magnification (Figs. 54, 55) in rats injected on days E15+16 and killed on day P60 indicates that the unlabeled cells of the anterolateral subnucleus are typically large neurons and that most of the small neurons are labeled throughout the paraventricular nucleus. However, there are many large neurons in all subdivisions of the nucleus that are also labeled (arrows in Figs. 54, 55).

The labeling pattern in a rat injected with [^{3}H]thymidine on day E15 and killed on day E22 (Fig. 56) closely resembles that seen in adults. The only component characterized by unlabeled cells is the anterolateral subnucleus. Rostrally (Fig. 56A), the unlabeled cells occupy a small dorsal rim but they occupy the entire dorsolateral protuberance more caudally (Fig. 56B). An attempt more

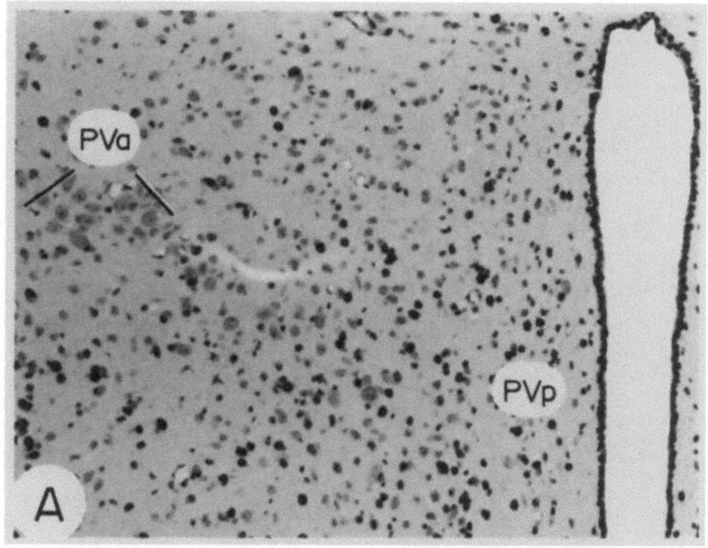

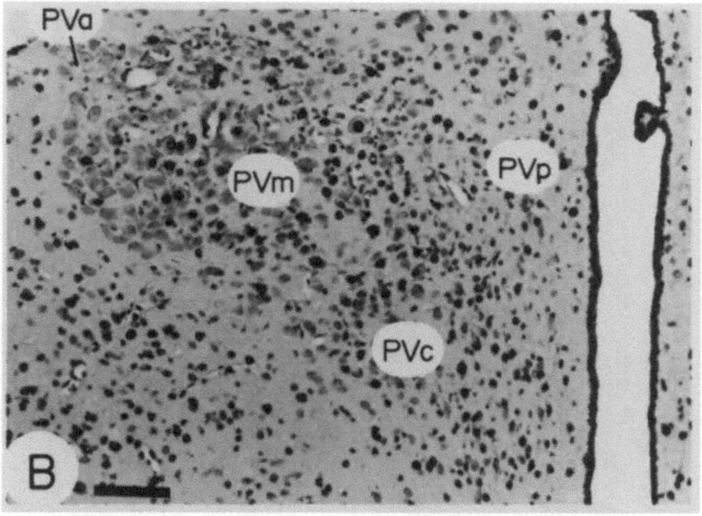

Figs. 52 and 53. Coronal thymidine radiograms, from rostral (Fig. 52 A) to caudal (Fig. 53 B) through the paraventricular nucleus, from adult rats labeled successively on days E16+17. Paraffin. *Scale*, 100 μm

clearly to subdivide the paraventricular nucleus in another rat injected on day E15 and killed on day E22 is made in Figs. 57 and 58. The neurons of the anterolateral subnucleus are mostly unlabeled, those of the medial subnucleus are heavily labeled, while in the posterior subnucleus many of the labeled cells have a relatively low grain concentration. In rats injected on day E16 and killed on day E22 (Fig. 59) only medially located parvocellular elements are labeled. We have suggested earlier that the neurons of the paraventricular nucleus originate in the germinal region composing the "diamond-shaped" ventricle (Altman and Bayer 1978 b: Figs. 7, 8). Our present observations indicate (arrows in Figs. 57, 58) that the cells of the parvocellular nucleus may derive from

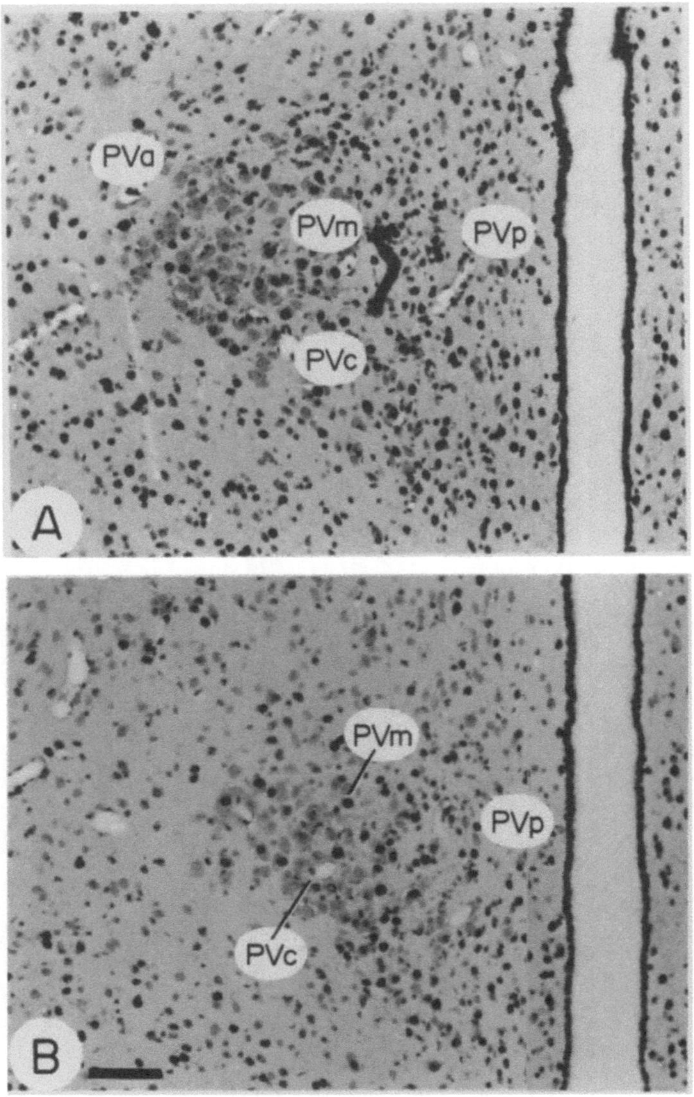

Fig. 53. Legend see Fig. 52

the neuroepithelium situated beneath the diamond-shaped ventricle, representing a portion of the third wave of neurons that form the periventricular field (Sect. 7.3).

Comments. The labeling pattern obtained here confirms our earlier suggestion (Altman and Bayer 1978a p. 958) that there is a dorsolateral-to-ventromedial gradient in the generation of neurons of the paraventricular nucleus. In light of the recently proposed parcellation of this nucleus (Armstrong et al. 1980; Swanson and Sawchenko 1983) it appears that the anterolateral subnucleus is composed of early-generated cells – those that could no longer be labeled by the morning of day E15, and presumably were generated on days E13 and

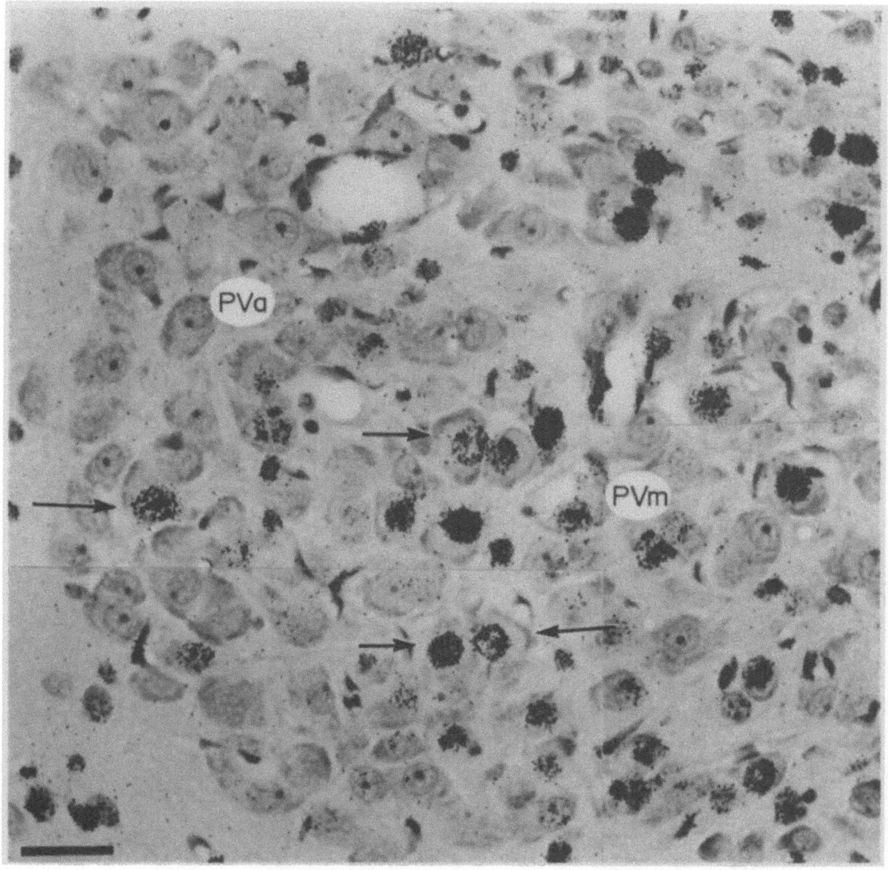

Figs. 54 and 55. High-power coronal thymidine radiograms, from rostral (Fig. 54) to caudal (Fig. 55), from an adult rat labeled on days E16+17, showing the cytology of labeled and unlabeled neurons in the paraventricular nucleus. Though most of the labeled neurons are small ones, some larger neurons (*arrows*) are also labeled. Paraffin. *Scale*, 30 µm

E14 (Altman and Bayer 1978a: Fig. 6). Most of the neurons of the medial and posterior subnuclei were labeled on day E15 but not on day E16; therefore, these cells were predominantly produced on day E15. The demonstration of neurophysin in the paraventricular nucleus by day E17 (Khachaturian and Sladek 1980; Whitnall et al. 1985) suggests that the synthesis of this protein associated with oxytocin and vasopressin begins in the paraventricular nucleus 1–2 days after the cessation of proliferation of its precursor cells.

The recently described segregation of oxytocin- and vasopressin-producing cells in the paraventricular nucleus raises the question whether the segregated early- and late-produced magnocellular neurons represent separate cell lines that differentiate into the two chemically distinct neurons. Because of discrepancies in the literature of the localization of oxytocin and vasopressin cells, however, this question cannot be resolved at the present. According to Rhodes et al. (1981), the medial subnucleus consists primarily of oxytocin cells, the lateral subnucleus is formed of a core of vasopressin cells with a rim of oxytocin cells, and the posterior subnucleus contains largely oxytocin cells. As the lateral

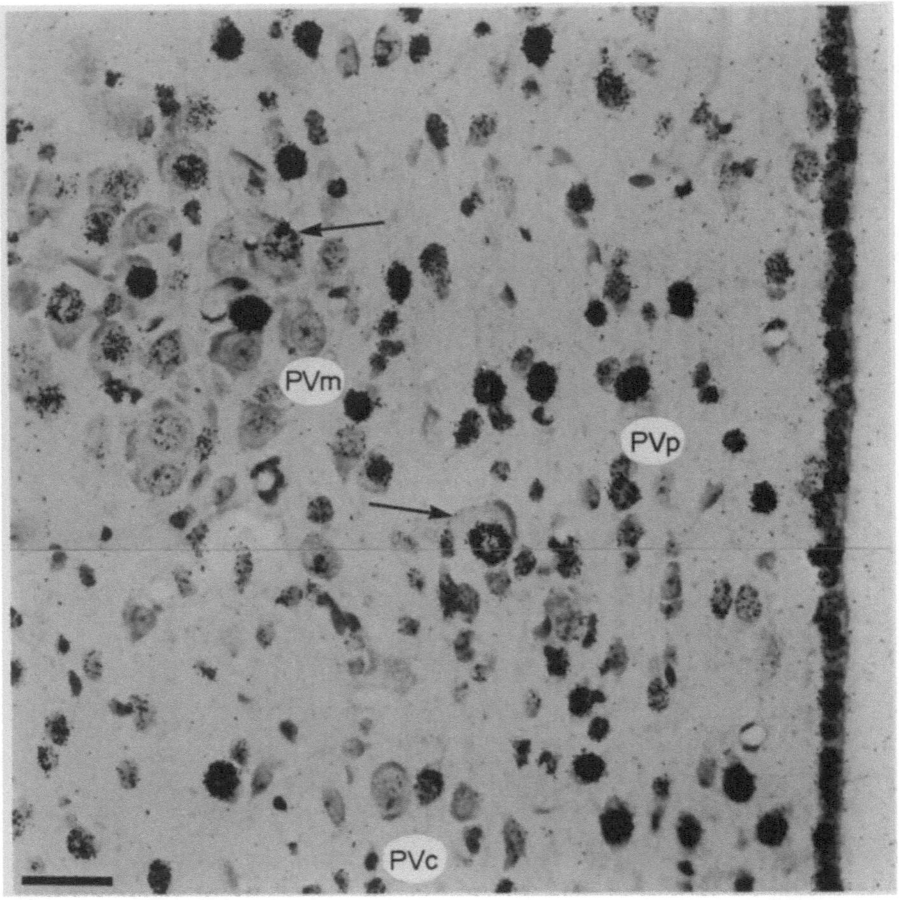

Fig. 55. Legend see Fig. 54

subnucleus is the only region containing vasopressin cells, this implies that the *vasopressin-producing* cells are generated *before* the oxytocin-producing cells. In contrast, Swanson and Sawchenko (1983 p. 280) conclude that the anterior and medial magnocellular subnuclei are composed almost exclusively of oxytocin cells and in the posterior subnucleus vasopressin cells are concentrated posterodorsolaterally. If vasopressin-staining cells are found only in the posterior subnucleus, this implies that *oxytocin-producing* cells are generated *before* the vasopressin-producing cells. Other published illustrations (Dierickx 1980 p. 140; Whitnall et al. 1985) fail to clarify this issue. We shall return to this problem in the next section, where we shall consider the differential localization of early- and late-produced neurons in the supraoptic nucleus in relation to the distribution of oxytocin- and vasopressin-staining cells in that structure.

6.2.3 The Supraoptic Nucleus

Background. The supraoptic nucleus is composed of large, intensely staining, and densely packed neurons. It is situated at the base of the anterior hypothala-

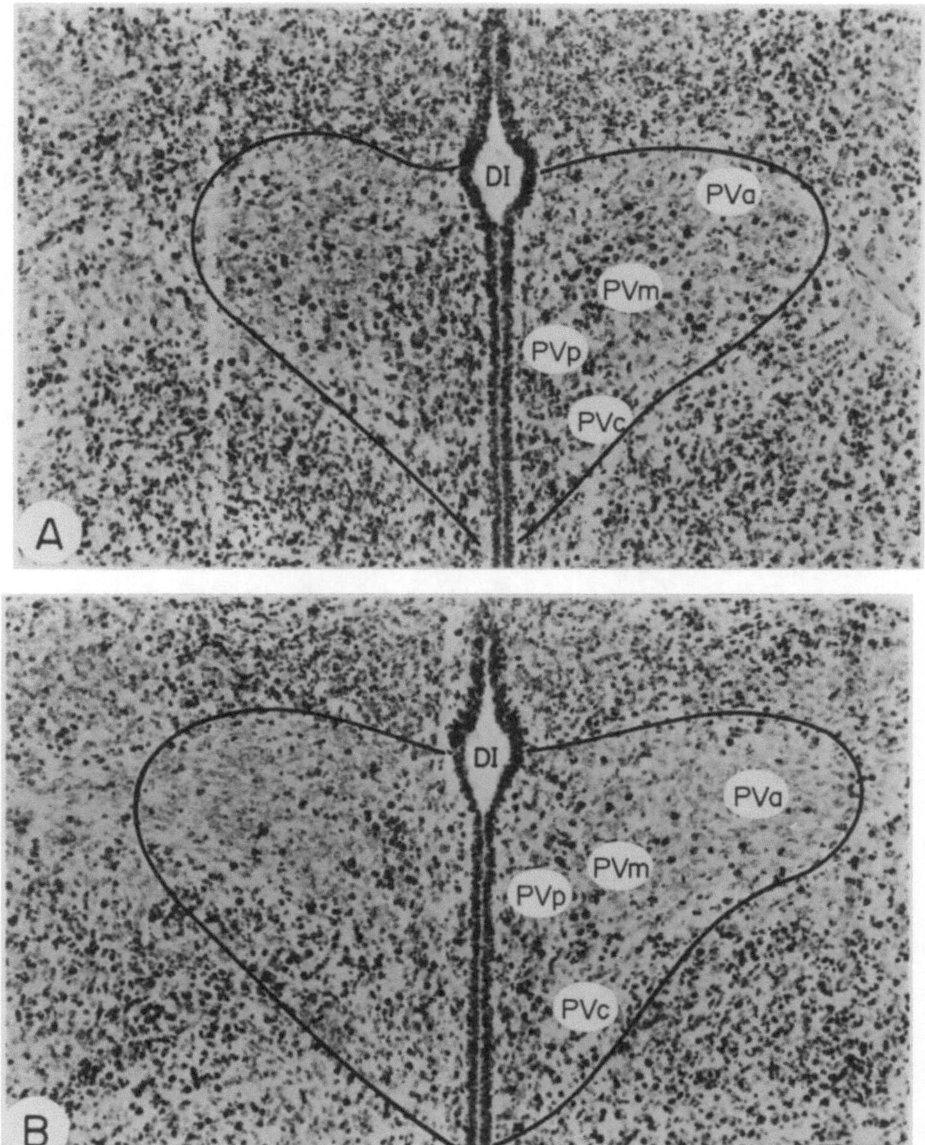

Fig. 56 A, B. Coronal thymidine radiograms, from rostral (**A**) to caudal (**B**) through the rostral part of the paraventricular nucleus, from a perinatal rat labeled on day E17. Paraffin. *Scale*, 100 μm

mus on the lateral border of the optic tract, extending from the optic chiasm anteromedially to the dorsally curving optic tract posterolaterally. Like the magnocellular neurosecretory cells of the paraventricular nucleus, the supraoptic neurons synthesize oxytocin, vasopressin, and neurophysin, and by way of the supraopticohypophysial tract convey these substances to the neural lobe of the pituitary gland (Bargmann 1949). Unlike the paraventricular nucleus, the supraoptic nucleus does not have a parvocellular component. In their immunocytochemical study, Rhodes et al. (1981) distinguished between a principal (ventro-

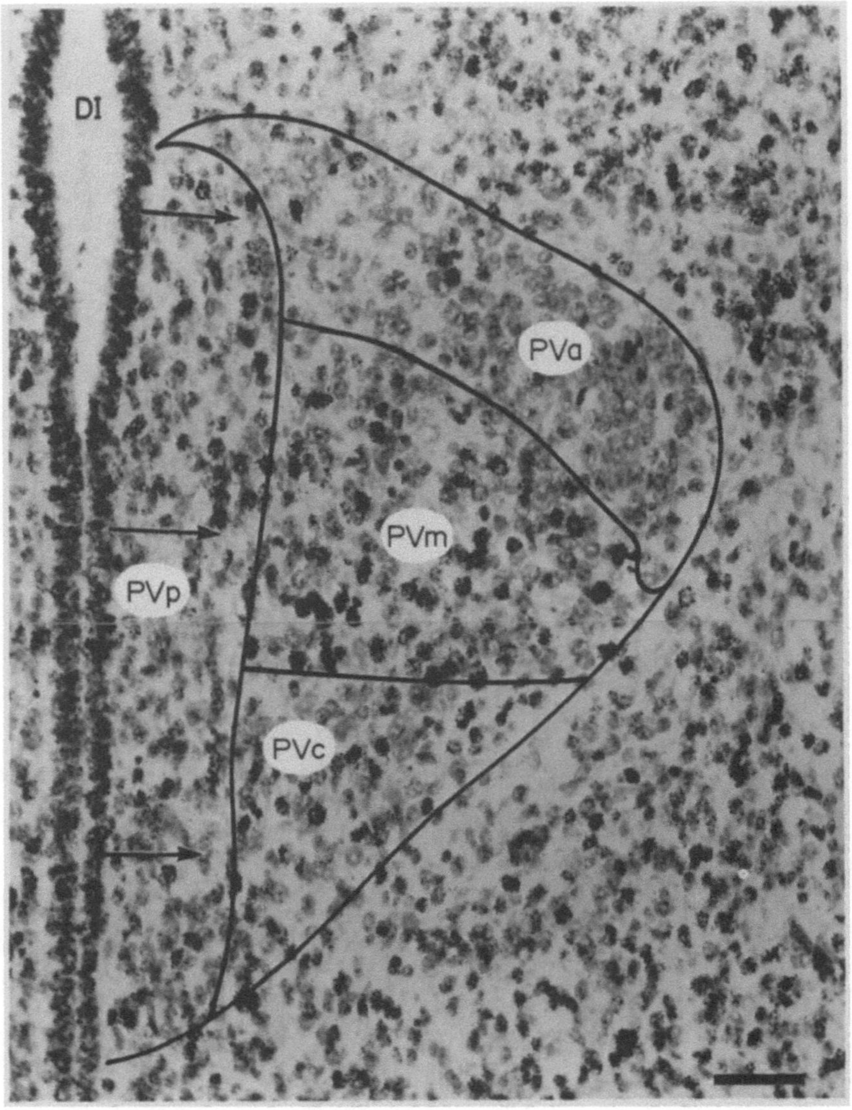

Figs. 57 and 58. High-power coronal thymidine radiograms, from rostral (Fig. 57) to caudal (Fig. 58) through the rostral part of the paraventricular nucleus, from a perinatal rat labeled on day E15. *Arrows* indicate the possible derivation of the parvocellular subnucleus from the neuroepithelium situated beneath the diamond-shaped ventricle. Paraffin. *Scale*, 50 μm

medial) and a retrochiasmatic (dorsolateral) component of the supraoptic nucleus, and claimed that in the principal component oxytocin cells predominate whereas in the retrochiasmatic region most cells stain for vasopressin.

Single-injection studies with [^{3}H]thymidine (Ifft 1972; Anderson 1978) indicate that heavily labeled supraoptic neurons are obtained in highest concentration with injection on day E14, and with reduced frequency on day E15. Our quantitative studies with double-labeling indicated that supraoptic neurons are produced at a nearly steady rate between days E13 and E15 (Altman and Bayer

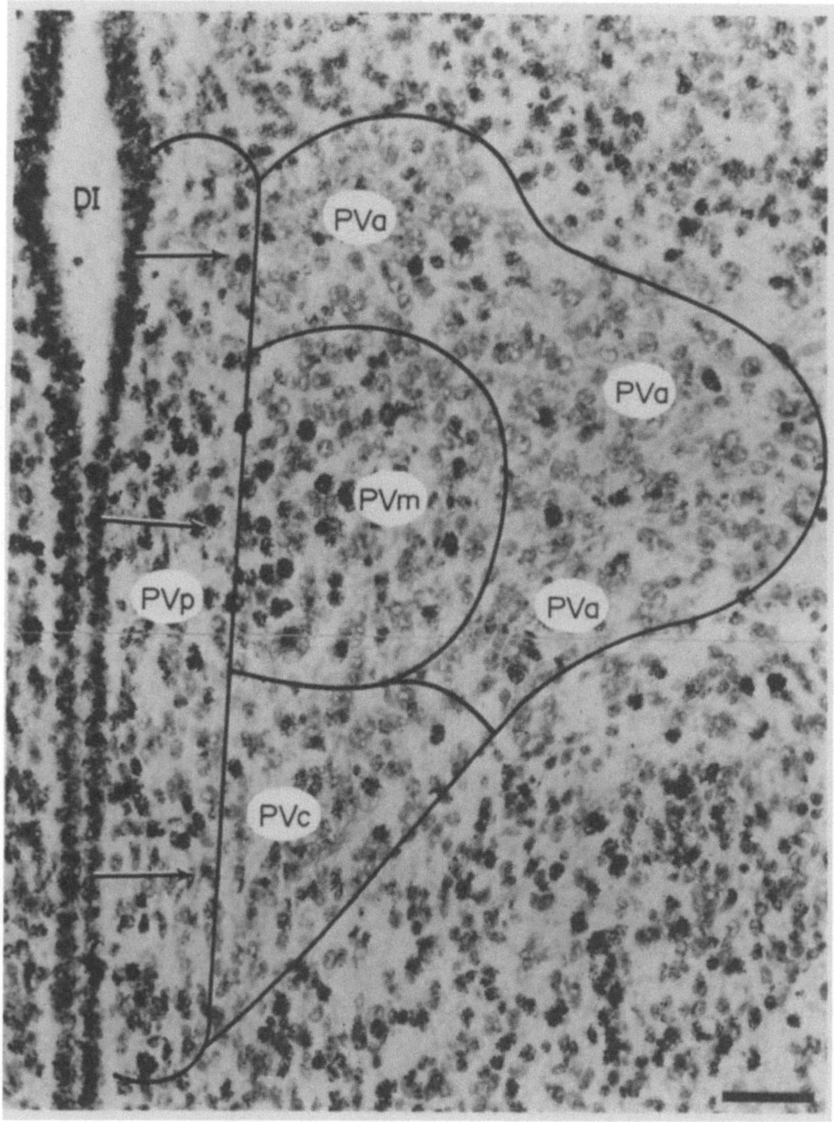

Fig. 58. Legend see Fig. 57

1978a: Fig. 5). We found a neurogenetic gradient in the supraoptic nucleus such that the earlier-generated neurons settle dorsolaterally and the later-generated neurons ventromedially (Altman and Bayer 1978a: Fig. 12). We postulated that the distally settling supraoptic neurons originate in the same "diamond-shaped" germinal matrix which is the source of the nearby-settling paraventricular neurons (Altman and Bayer 1978b: Figs. 7, 8). The distally settling magnocellular neurons of the supraoptic nucleus begin their cytological differentiation much earlier than the proximal paraventricular neurons. According to Wolf and Trautman (1977), neurophysin-containing cells can be seen in the supraoptic nucleus as early as day E16 but not until day E19 in the paraventricular nucleus. Oxytocin- and vasopressin-associated neurophysins were, likewise, detected by

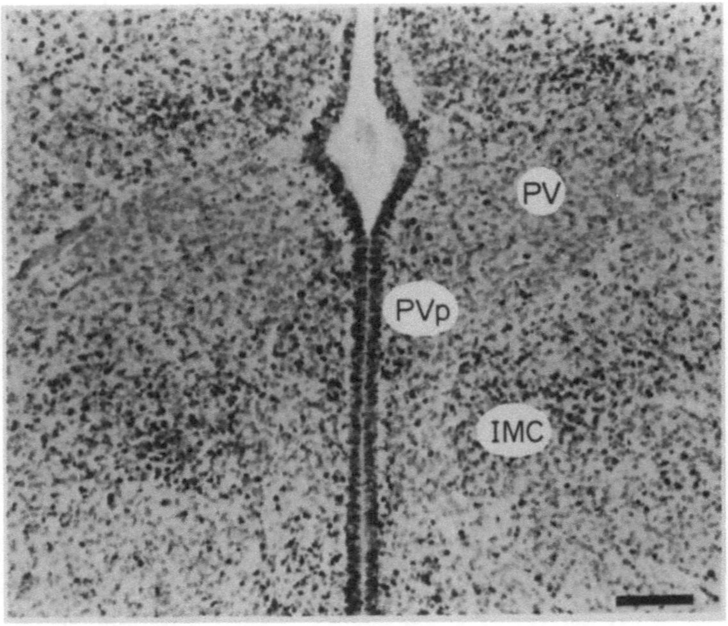

Fig. 59. The continued labeling of cells of the parvocellular subnucleus and the internuclear magnocellular neurons in a perinatal rat tagged on day E16. Paraffin. *Scale*, 100 μm

Whitnall et al. (1985) by day E16 in the paraventricular nucleus but their illustrations also indicate the early appearance (days E16 and E17) of these cells in the paraventricular nucleus.

The term "internuclear magnocellular neurosecretory neurons" refers to a collection of large neurosecretory cells scattered in a broad trajectory between the paraventricular nucleus and the supraoptic nucleus. Some of these cells form clusters around larger blood vessels and one of these clusters has been designated as the nucleus circularis (Peterson 1966). Some of these cells, but not all, are labeled with injections on day E15 (Altman and Bayer 1978a: Fig. 14), suggesting a time of origin roughly paralleling the neurons of the paraventricular and supraoptic nuclei.

Results. In rats injected with [^{3}H]thymidine on day E14 and killed on day E20 (Fig. 60) a high proportion of the supraoptic neurons are labeled. In rats injected on day E15 and killed on day E20 (Fig. 61) the supraoptic cells are labeled in the juxtachiasmatic ventromedial (more rostral) region but they tend to be unlabeled in the retrochiasmatic dorsolateral (more posterior) region. The same pattern was obtained in adult rats injected on days E15+16 (Fig. 62). In adult rats the anterior location of the late-produced ventromedial neurons (Fig. 62A) and the posterior location of the early-produced dorsolateral neurons (Fig. 62C) is more pronounced than in fetal rats (Fig. 61).

The supraoptic nucleus is not clearly identifiable on day E15 but the earliest cells may begin to settle on this day, as indicated in the radiogram of the rat injected on day E14 and killed on day E15 (SO? in Fig. 63). Scattered between the putative source of the secretory magnocellular neurons (mh in

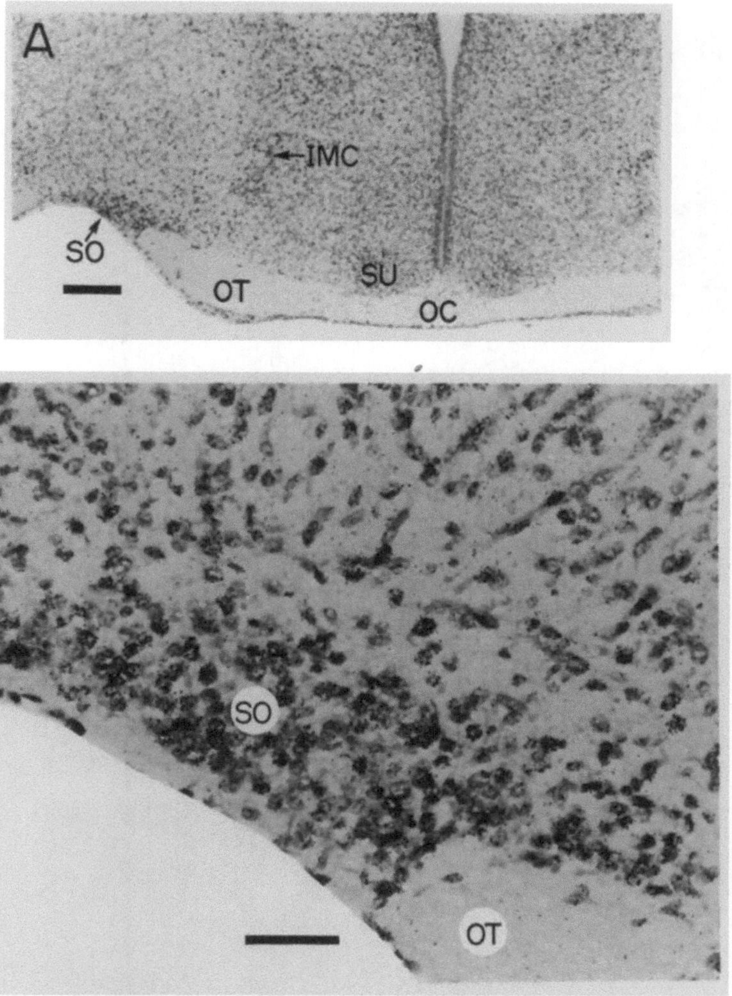

Fig. 60. A The labeling pattern of the supraoptic nucleus in a perinatal rat tagged on day E14. **B** Detail. Paraffin. *Scales*, **A** 200 µm; **B** 50 µm

Fig. 63) and the future supraoptic nucleus are labeled cells that may be migrating supraoptic cells (mi). In rats injected on day E14 and killed on day E16 (Fig. 64) a proportion of the labeled cells have settled in the formative supraoptic nucleus. In rats injected on day E15 and killed the next day (Fig. 65A) the earliest-settling dorsolateral cells are not labeled; these were evidently produced before day E15. In rats injected on day E15 and killed 2 days later (Fig. 65B) labeled cells begin to approximate the ventromedial region of the supraoptic nucleus. In rats injected on day E15 and killed 3 days later (Fig. 65C) all the cells generated on day E15 appear to have settled in the ventromedial part of the supraoptic nucleus. The dorsolateral-to-ventromedial settling pattern of supraoptic cells produced between days E13 and E15 is illustrated in Fig. 66. In the rat injected on day E14 and killed 2 days later, the dorsolaterally situated neurons are unlabeled; those situated more ventromedially are labeled (Fig. 66A). The unlabeled dorsolateral cells were presumably generated on day E13, the adjacent

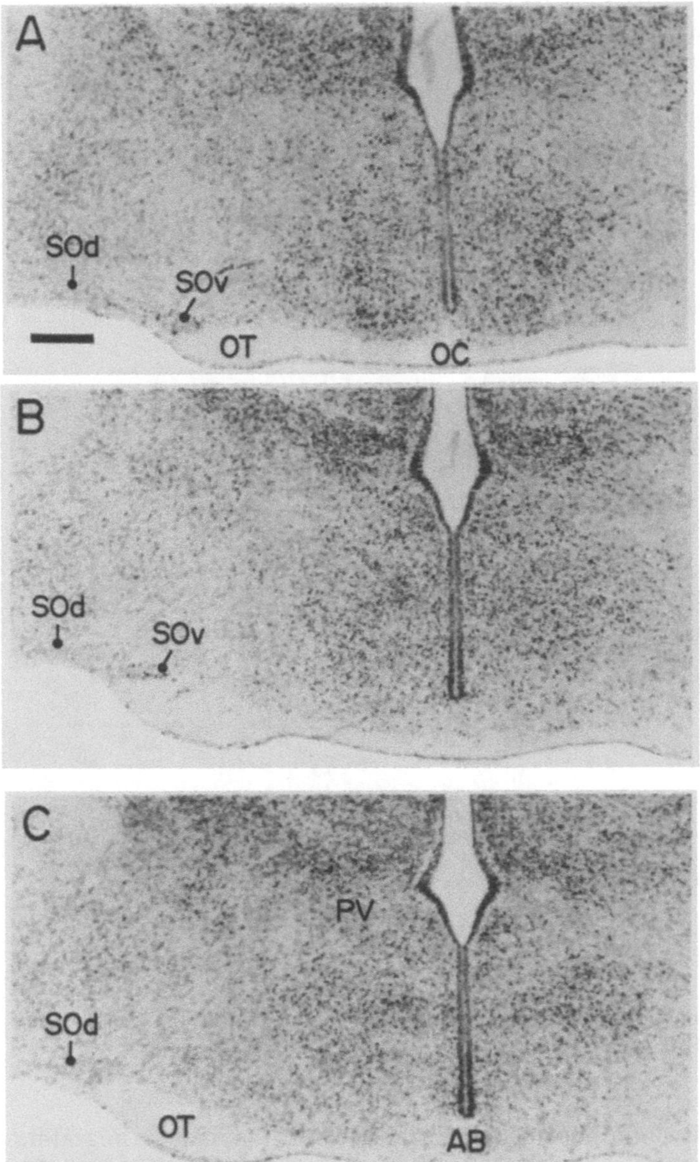

Fig. 61 A–C. Coronal thymidine radiograms, from rostral (**A**) to caudal (**C**), from a perinatal rat tagged on day E15, showing the dorsolateral-to-ventromedial neurogenetic gradient in the labeling of supraoptic neurons. Paraffin. *Scale*, 200 µm

labeled cells on day E14. In rats labeled on day E15 and killed 2 days later (Fig. 66B) the supraoptic nucleus has become extended ventromedially and only those cells situated in the latter region are labeled.

The majority of the neurons of the internuclear magnocellular neurons, specifically those of the circular nucleus (IMC in Fig. 67), are labeled in rats injected on day E15 and killed on day E22. Most of these cells are still labeled with injections made on day E16, indicating that, in this respect, they resemble the

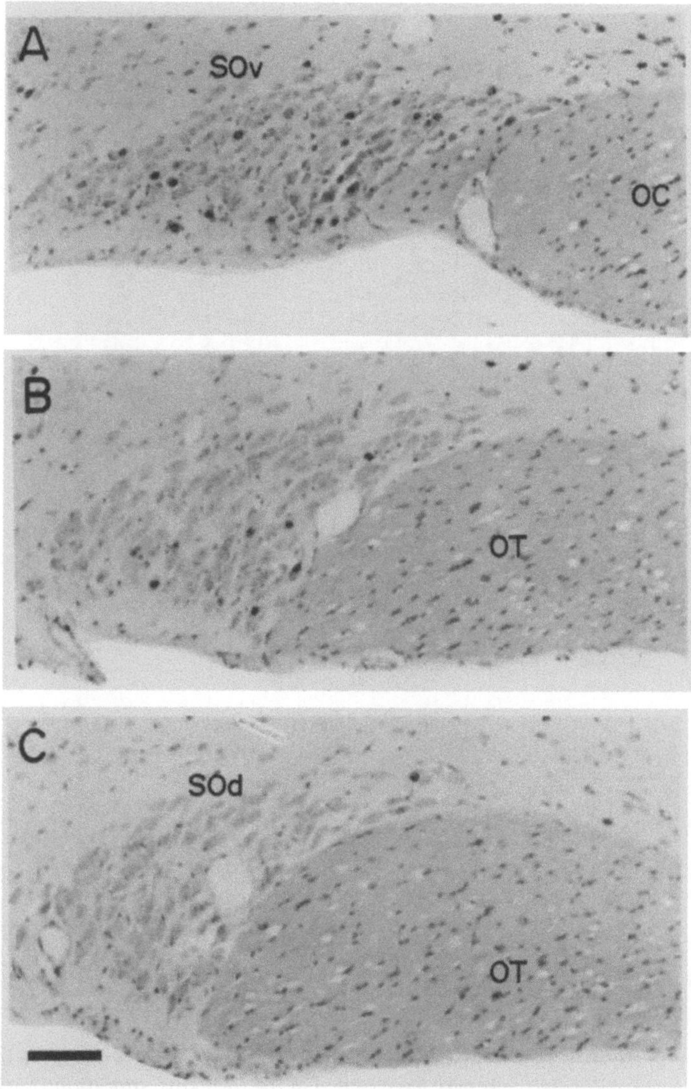

Fig. 62 A–C. Coronal thymidine radiograms, from rostral (**A**) to caudal (**C**), from an adult rat labeled on days E15 + 16. Paraffin. *Scale*, 100 μm

late-generated cells of the parvocellular component of the paraventricular nucleus (Fig. 59).

Comments. The present study supports our earlier quantitative result that the supraoptic neurons are produced over a 3-day period between days E13 and E15 (Altman and Bayer 1978a: Fig. 5). The migration of the cells takes 2–3 days: the cells generated on day E13 begin to settle in small numbers on day E16 (Fig. 66A) and the settling of cells produced on day E15 appears to be completed by day E18 (Fig. 65C). The settling pattern of supraoptic neurons follows a gradient along the optic tract from dorsolateral (retrochiasmatic) to ventromedial (principal). The settling in the vicinity of the optic tract raises

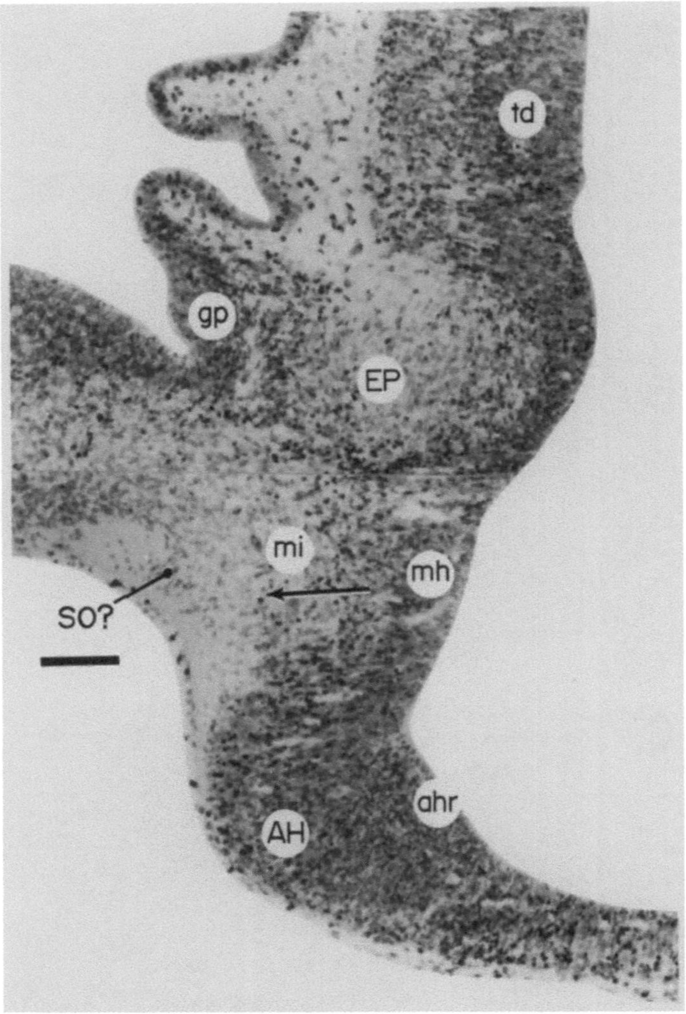

Fig. 63. Coronal thymidine radiogram from a rat tagged on day E14 and killed on day E15. A few unlabeled cells may have settled in the formative supraoptic nucleus; others are presumed to be migrating (*arrow*) from the magnocellular hypothalamic neuroepithelium. Paraffin. *Scale*, 100 μm

the possibility that optic fibers exert some guiding influence on the migrating supraoptic cells. But this is ruled out by our observations. Optic fibers do not reach the chiasmatic region until day E16 and the ingrowth of fibers along the ventrolateral border of the hypothalamus is not in progress until day E17. But some supraoptic neurons begin to settle as early as day E15 (Fig. 63) and by day E16 the supraoptic nucleus is quite prominent (Figs. 64, 66A). The fiber bundle that is recognizable near the supraoptic nucleus (which is initially situated more laterally than in mature animals) is presumed to be the supraoptic decussation (SD in Fig. 64). The settling of supraoptic neurons in a dorsolateral-to-ventromedial order can be attributed to the orderly migration of cells from dorsal-to-ventral in the chronological sequence of their production (Fig. 66).

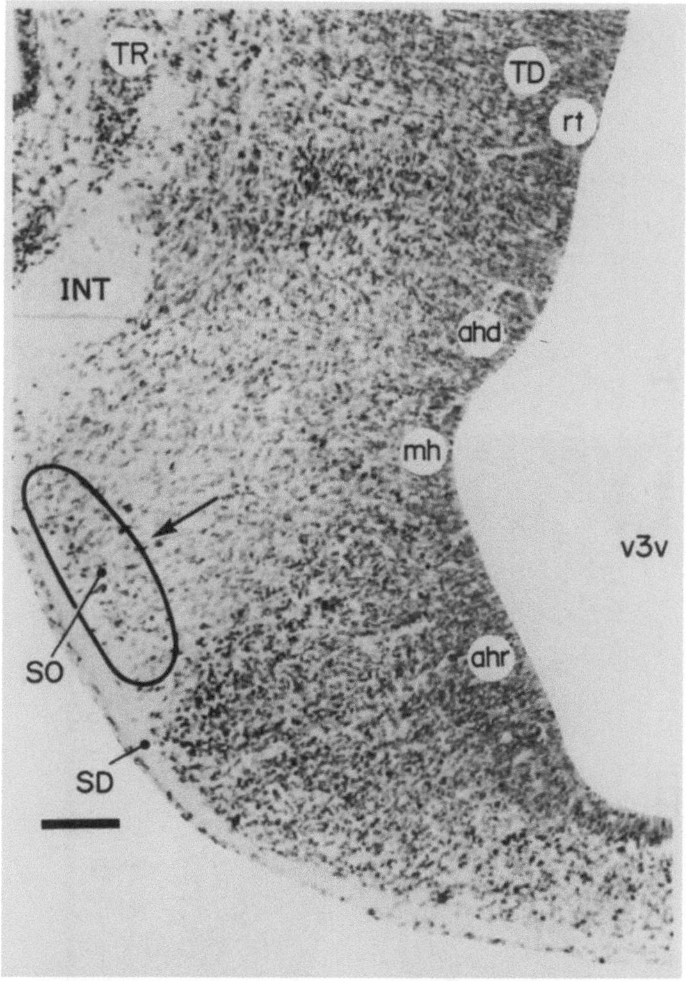

Fig. 64. Coronal thymidine radiogram from a rat tagged on day E14 and killed on day E16. Labeled and unlabeled cells are settling in the supraoptic nucleus, which in young embryos is situated in a lateral position. Paraffin. *Scale*, 100 µm

This pattern suggests the possibility that in the magnocellular germinal matrix (mh) the gradient of neurogenesis is from dorsal to ventral (vertical arrows in Fig. 66).

Our present observations are compatible with the hypothesis that the neurophysin-associated, oxytocin-, and vasopressin-synthesizing magnocellular neurons of the hypothalamus all derive from a single neuroepithelial source, what we have earlier designated as the "diamond-shaped ventricle" (Altman and Bayer 1978b). In fact, at the time that the magnocellular secretory neurons are produced this expansive ventricular region undergoes appreciable configurational changes (Figs. 63, 64). On day E16 (Fig. 65A) it becomes hexagonal in shape and by day E18 it assumes a diamond shape (Fig. 65C). Interestingly, the order of cellular maturation of the magnocellular neurons is the reverse of what might be expected. The supraoptic and paraventricular neurons are

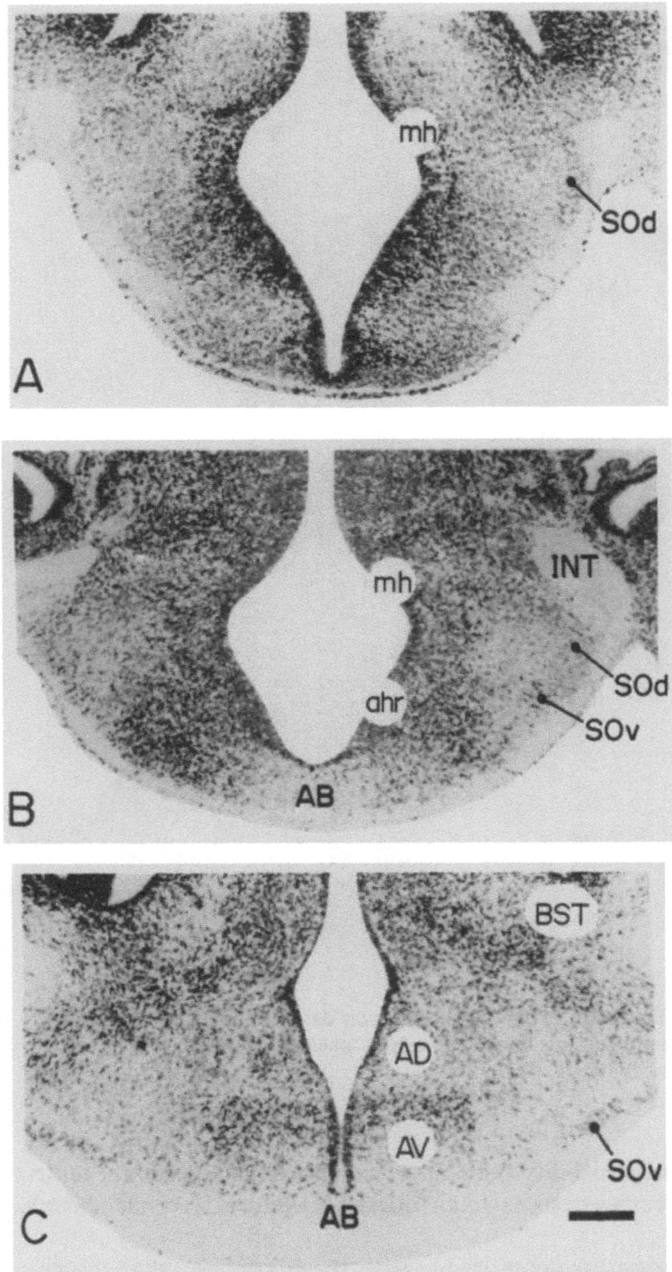

Fig. 65 A–C. Coronal thymidine radiograms from rats that were labeled on day E15 and killed on day E16 (**A**), E17 (**B**), and E18 (**C**). The earliest settling dorsolateral cells are unlabeled 1 day after injection (**A**). Some labeled cells appear ventromedially 2 days after injection (**B**), and 3 days after injection (**C**) a high proportion of the ventromedial neurons are labeled. Paraffin. *Scale*, 200 μm

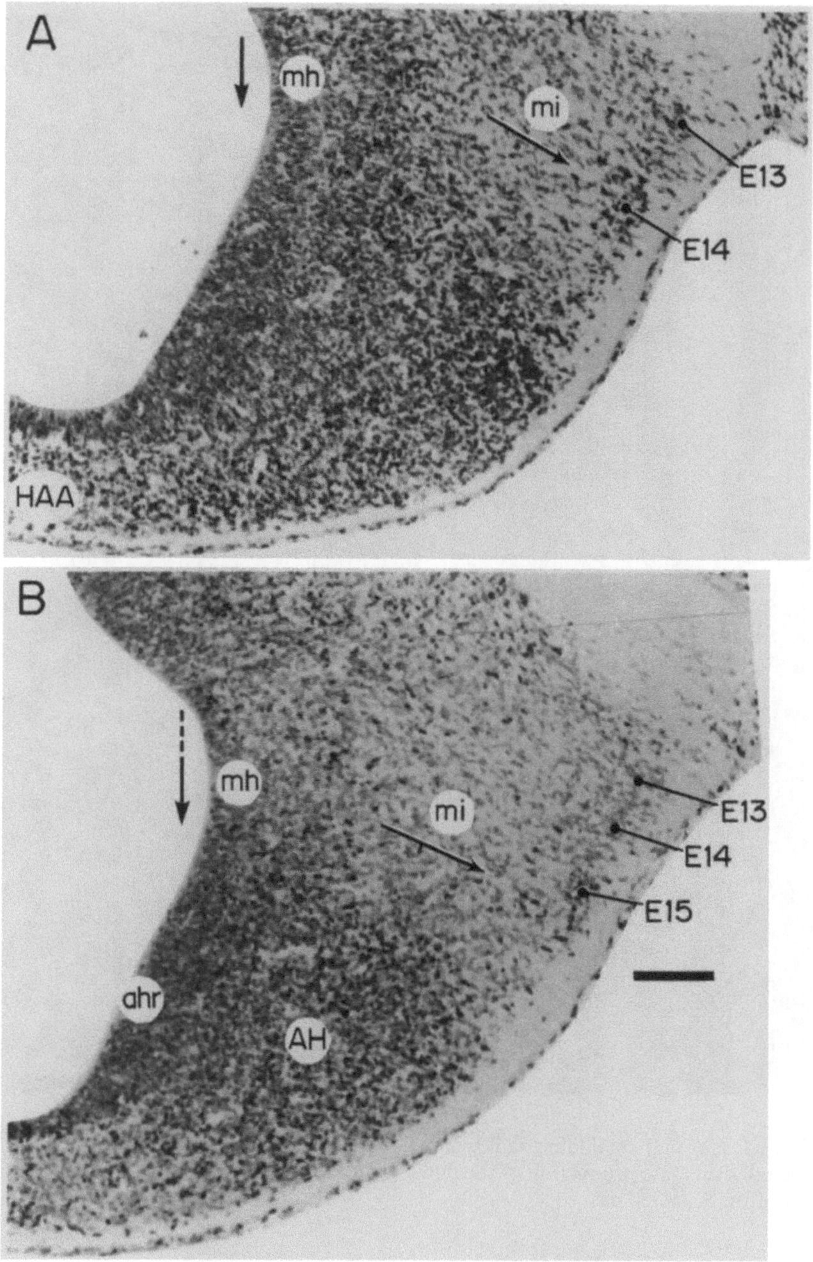

Fig. 66. A Coronal thymidine radiogram from a rat tagged on day E14 and killed on day E16. The dorsolaterally settling neurons are unlabeled. These are assumed to have been generated on day E13. The adjacent cells situated more ventromedially are labeled; these are presumed to have been produced on day E14. **B** Radiogram from a rat injected on day E15 and killed on day E17. Only the most ventromedially situated cells are labeled. This is the complement which was apparently generated on day E15. *Radial arrows in* **A** *and* **B** indicate the presumed migratory path of the supraoptic neurons. *Horizontal arrows* suggest the possible displacement of the active magnocellular hypothalamic neuroepithelium from dorsal to ventral. Paraffin. *Scale*, 100 μm

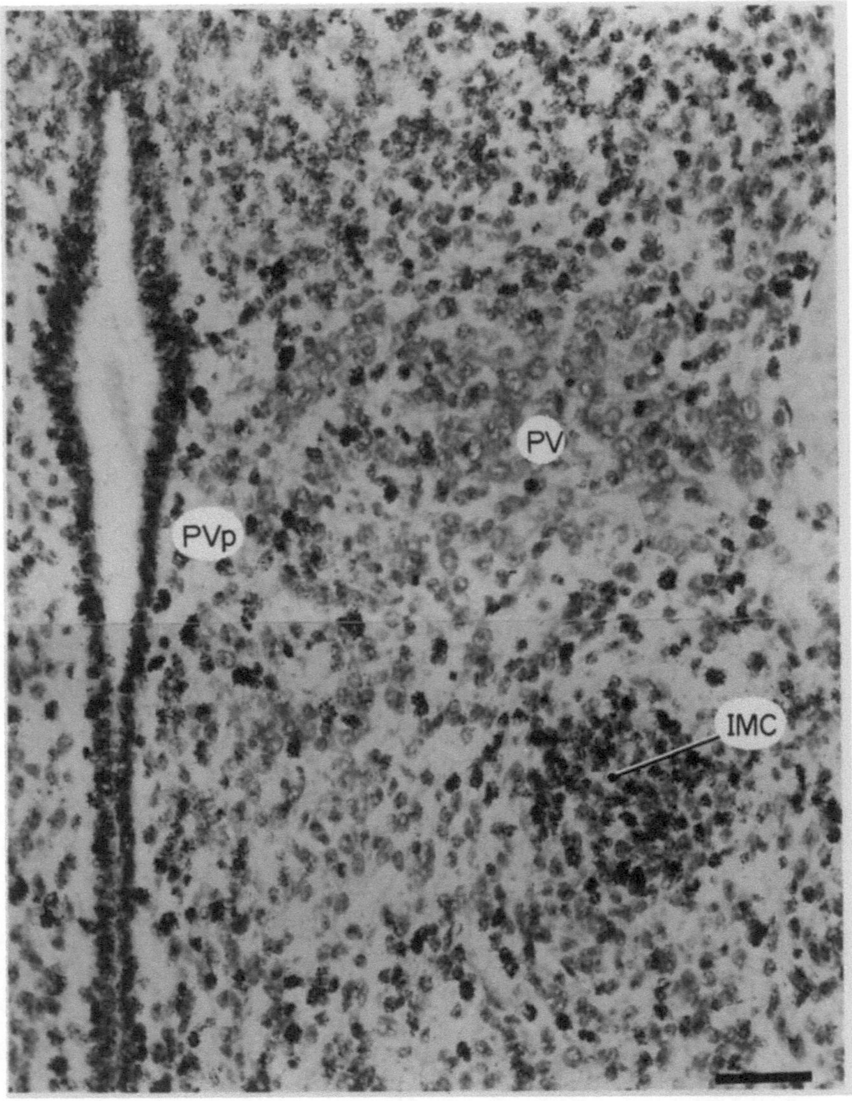

Fig. 67. Coronal thymidine radiogram from a perinatal rat labeled on day E15 to illustrate the relatively late production of the circular nucleus of the internuclear magnocellular complex. Paraffin. *Scale*, 50 μm

generated concurrently (Altman and Bayer 1978a Figs. 5, 6) but the distally settling supraoptic neurons mature earlier than the proximally settling paraventricular neurons. It is possible that the maturation of the magnocellular secretory neurons is associated with the growth of blood vessels and capillaries from the pial surface inward.

The description (Rhodes et al. 1981) that in the dorsolateral (retrochiasmatic) supraoptic nucleus most neurons stain for vasopressin whereas in the ventromedial (principal or juxtachiasmatic) supraoptic nucleus most cells stain for oxytocin implies, when combined with our results, that the vasopressin-

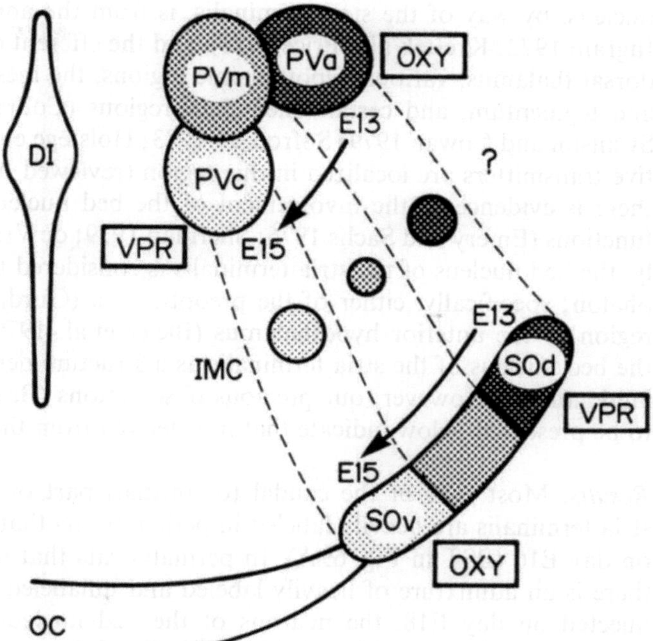

Fig. 68. Schematic illustration of the conflicting neurogenetic prediction (*broken lines*) and available reports of the distribution of oxytocin- and vasopressin-synthesizing neurons in the paraventricular nucleus and the supraoptic nucleus. See text for details. The illustrated chronological distribution of the internuclear magnocellular neurons is strictly hypothetical

synthesizing neurons are generated earlier than the oxytocin-synthesizing cells (Fig. 68). This could mean that the vasopressin cells derive from a line of stem cells located in the dorsal part of the magnocellular neuroepithelium and the oxytocin cells from a cell line located in its ventral part. But if this were so, it would follow that the same pattern holds also for the proximally settling paraventricular neurons (broken lines in Fig. 68), namely, that the earlier-generated neurons of the anterolateral subnucleus are vasopressin-producing cells, and the later-generated neurons of the posterior subnucleus are oxytocin-producing cells. But many workers (e.g., Dierickx 1980; Swanson and Sawchenko 1983) report the opposite pattern (Fig. 68). Therefore, the question must remain open whether or not the time and site of origin of magnocellular secretory neurons predetermines if they will produce oxytocin or vasopressin.

6.2.4 The Problem of the Bed Nucleus of the Stria Terminalis

Background. The bed nucleus of the stria terminalis is situated rostrally between the ventrolateral aspect of the septum and the anterior commissure, that is, in the telencephalon. Caudally it extends to the preoptic area and as far into the body of the diencephalon as the rostral margin of the anterior hypothalamus. Its rostral part is sometimes referred to as the medial bed nucleus, and its caudal part as the lateral bed nucleus. The major afferent supply of the bed nucleus, by way of the stria terminalis, is from the amygdala (De Olmos and Ingram 1972; Krettek and Price 1978), and the efferent outflow is to the mediodorsal thalamus, various hypothalamic regions, the mesencephalic central gray and tegmentum, and certain medullary regions (Conrad and Pfaff 1976a, b; Swanson and Cowan 1979; Sofroniew 1983; Holstege et al. 1985). Various putative transmitters are localized in this region (reviewed by Sofroniew 1983), and there is evidence of the involvement of the bed nucleus in male reproductive functions (Emery and Sachs 1976; Sheridan 1979; de Vries et al. 1985). Generally, the bed nucleus of the stria terminalis is considered to be part of the diencephalon; specifically, either of the preoptic area (Gurdjian 1927) or the dorsal region of the anterior hypothalamus (Bleier et al. 1979). Bleier et al. describe the bed nucleus of the stria terminalis as a structure derived from the hypothalamic anlage. However, our previous observations (Bayer 1979a, b) and those to be presented below indicate that it is derived from the lateral ventricle.

Results. Most cells of the caudal (or medial) part of the bed nucleus of the stria terminalis are heavily labeled in perinatal rats that received [^{3}H]thymidine on day E16 (BST in Fig. 69A). In perinatal rats that were tagged on day E17 there is an admixture of heavily labeled and unlabeled cells (Fig. 69B). In rats injected on day E18, the neurons of the bed nucleus are no longer labeled (not shown). This establishes that the bulk of the neurons of the caudal part of the bed nucleus are generated on day E16 and a few are still being generated on day E17, confirming our earlier quantitative study (Bayer 1979a). In rats tagged on day E16 (Fig. 69A), the boundary between the bed nucleus and the posterodorsal preoptic region is not evident because both are composed of heavily labeled cells. However, in rats injected on day E17 (Fig. 69B), the neurons of the posterodorsal preoptic region are no longer labeled dorsolaterally, while in the contiguous bed nucleus there are scattered heavily labeled cells, permitting the delineation of the boundary between the two structures.

The impression gained from radiograms in perinatal rats is that unlike the posterodorsal preoptic region, which is derived from the third ventricle neuroepithelium (Sect. 6.1), the bed nucleus of the stria terminalis is derived from the neuroepithelium of the lateral ventricle, specifically from the base of its medial horn (vlm in Fig. 69A, B). This telencephalic origin of the caudal portion of the bed nucleus is supported by observations in young embryos. In rats injected on day E16 and killed 1 day later on day E17 (Fig. 70A), the presumed germinal source of the bed nucleus (bst) is surrounded by differentiating, heavily labeled cells, some of them apparently moving ventromedially. In rats injected on day E15 and killed 2 days later on day E17 (Fig. 70B), the labeled cells have apparently streamed much farther into the body of the diencephalon. This suggests that, like the rostral portion of the bed nucleus of the stria terminalis, which is unquestionably of telencephalic origin, so also its caudal part derives from the lateral ventricle (Fig. 11) and not from the third ventricle.

Comments. These observations strongly indicate that the caudal (medial) part of the bed nucleus of the stria terminalis, although it is located unquestionably within the body of the diencephalon, is of telencephalic derivation. While its boundary with the posterodorsal preoptic region is not easy to discern in normal material, the two can be distinguished in radiograms because of differences

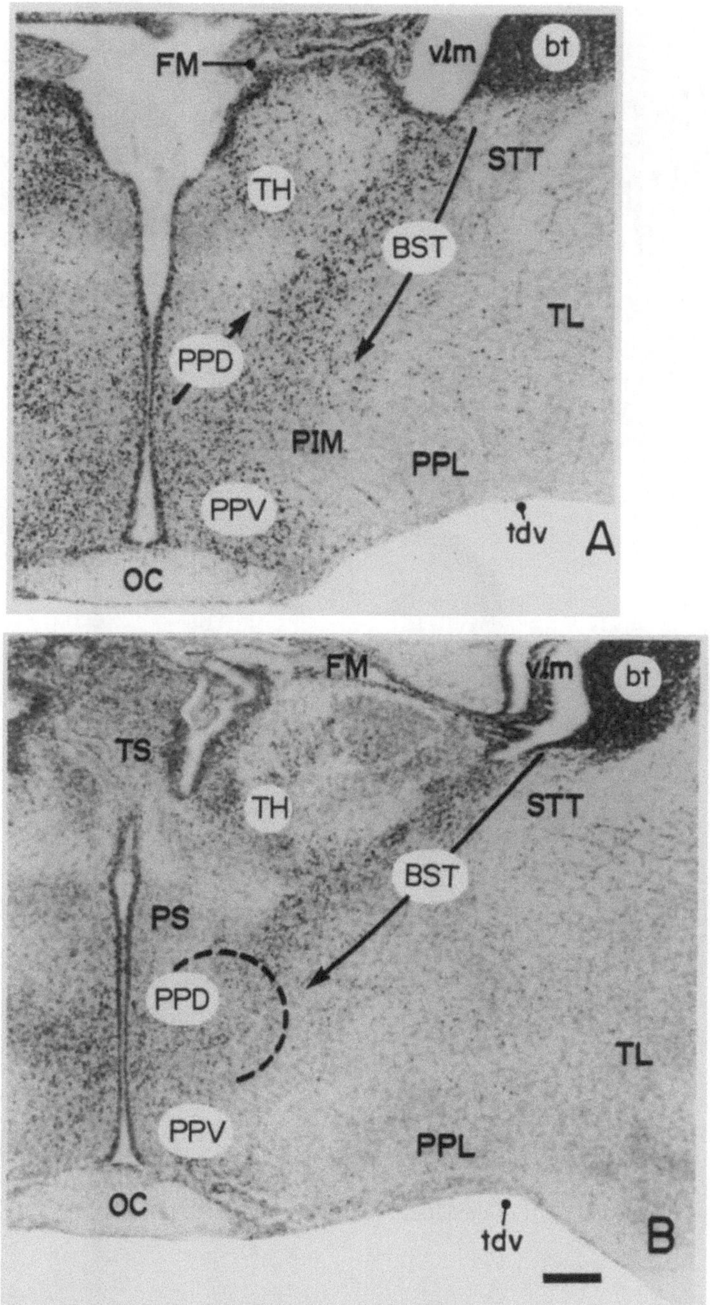

Fig. 69. A Coronal thymidine radiogram from a perinatal rat labeled on day E16. Most neurons of the caudal (medial) portion of the bed nucleus of the stria terminalis are labeled, and so are also the neurons of the posterodorsal preoptic region. *Arrows* indicate the presumed derivation of the former from the lateral ventricle neuroepithelium, and of the latter from the third ventricle neuroepithelium. **B** Thymidine radiogram from a perinatal rat labeled on day E17. In the bed nucleus of the stria terminalis there are still many labeled cells but only few in the posterodorsal preoptic region. This allows the delineation of the approximate boundary of the latter (*broken semicircle*) and of the bed nucleus of the stria terminalis (*arrow*). Paraffin. *Scale*, 200 μm

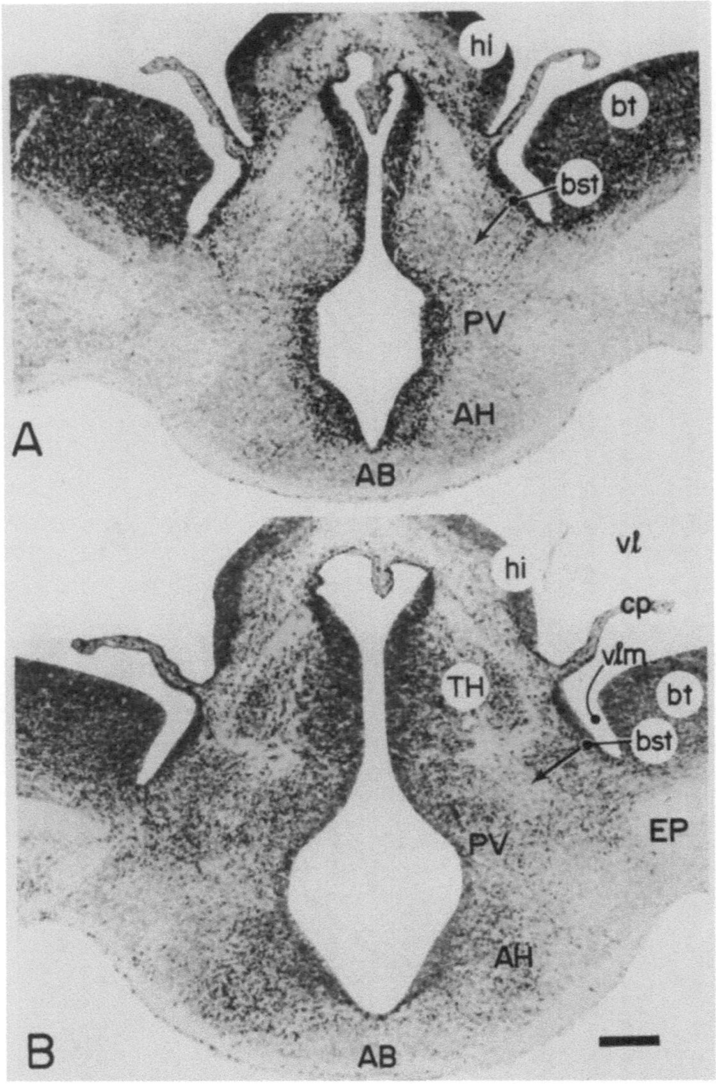

Fig. 70 A, B. Coronal thymidine radiograms from day-E17 rats, one tagged 1 day earlier (**A**), the other 2 days earlier (**B**). The inferred origin of the ventromedially directed stream of neurons of the caudal bed nucleus of the stria terminalis (Fig. 69) in the basal neuroepithelium of the medial horn of the lateral ventricle is indicated (*arrows*). Paraffin. *Scale*, 200 μm

in the time of origin of their neurons. The two, if our interpretation is correct, differ in their germinal origins, one deriving from the neuroepithelium of the third ventricle, and the other from the lateral ventricle. While, in general, the derivation of diencephalic structures from the third ventricle and of telencephalic structures from the lateral ventricle is well established, the caudal part of the bed nucleus of the stria terminalis is an exception to this. In this general region the boundary between the diencephalon and the telencephalon is apparently not strictly maintained. It may be recalled that the neurons of the entopeduncular nucleus, which we have traced to the third ventricle neuroepithelium (Sect. 5.2), settle within the telencephalon.

6.3 Development of the Anterior Hypothalamus: Caudal Region

6.3.1 The Ventromedial Nucleus

Background. The ventromedial hypothalamic nucleus has received considerable attention because of its apparent involvement in a variety of physiological and behavioral regulations. These include: appetite and feeding (Hetherington and Ranson 1942; Brobeck et al. 1943; Grossman 1976); gastric secretion (Ridley and Brooks 1965); the release of growth hormone, insulin, and glucagon (Frohman and Bernardis 1971; Frohman et al. 1974), and of thyrotropin (Brownstein et al. 1974), corticotropin (Szentágothai et al. 1968), and gonadotropins (Palkovits et al. 1974); and reproductive functions (Kennedy 1963; Matthews and Edwards 1977; Barfield and Chen 1977; Pfaff and Sakuma 1979). The mechanisms whereby these multifarious functions are influenced by ventromedial neurons are not known. These neurons have long dendrites that extend into adjacent hypothalamic regions, and the capsule that surrounds the nucleus is made up partly of afferents (Millhouse 1973) from a variety of structures, including the septum, nucleus accumbens, amygdala, globus pallidus, medial preoptic nucleus, and several hypothalamic nuclei (Conrad and Pfaff 1976a; Kita and Oomura 1982). The efferent projections of the ventromedial nucleus are, anteriorly, to the septum, amygdala, the bed nucleus of stria terminalis and preoptic area, and, posteriorly, to the posterior hypothalamus, the mammillary complex, the ventral tegmental area, and the central gray (Saper et al. 1976; Palkovits and Záborszky 1979). The ventromedial nucleus has at least two components, the pars ventralis situated ventrolaterally and the pars dorsalis situated dorsomedially. The neurons of the ventromedial nucleus are produced between days E13 and E17 (Altman and Bayer 1978a: Fig. 7) with a gradient from ventrolateral to dorsomedial (Ifft 1972; Altman and Bayer 1978a).

Results. In fetal rats injected with [^{3}H]thymidine on day E13 (Fig. 71 A) all the neurons of the ventromedial nucleus are labeled. Those located in the pars ventralis (VMv) tend to be heavily labeled, while those situated in the pars dorsalis (VMd) tend to be lightly labeled. In fetal rats injected on day E14 many of the neurons of the pars ventralis are no longer labeled (arrows in Fig. 71 B), indicating that these cells are produced some time between the mornings on day E13 and E14. With injection delayed until day E15 (Fig. 72A) most of the neurons of the pars dorsalis of the ventromedial nucleus are no longer labeled, with the exception of a strip of cells overlying the arcuate nucleus. Many of these cells are still labeled following injections on day E16 (Fig. 72B) and day E17 (Fig. 72C). Because of its discrete labeling chronology, we designate this region as the pars basalis or *basal subnucleus of the ventromedial nucleus* (VMb). The location of this subnucleus, its distinctiveness from the arcuate nucleus (AR), and its continuity with the anterior periventricular area (AP) are illustrated in a radiogram from an adult rat injected with [^{3}H]thymidine on days E16+17 (Fig. 73).

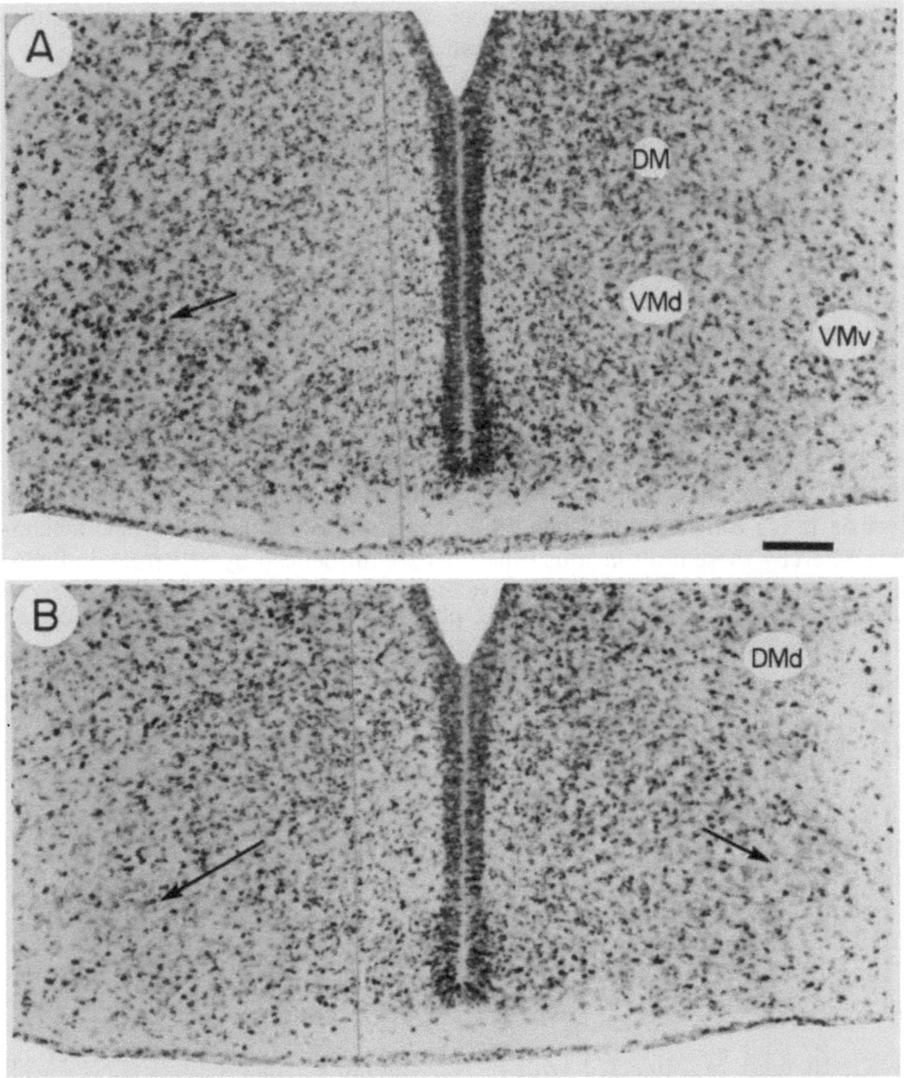

Fig. 71 A, B. Coronal thymidine radiograms from perinatal rats labeled on day E13 (**A**) and day E14 (**B**). In rats injected on day E13 heavily labeled cells are limited to the pars ventralis of the ventromedial nucleus (**A**). Many of these cells are no longer labeled in the rat injected on day E14 (*arrows in* **B**). Paraffin. *Scale*, 100 μm

As early as day E16 (Fig. 74A), and possibly day E15 (Fig. 75A), the settling of ventromedial neurons is recognizable. The likely neuroepithelial source of these neurons is the lower half of the ventral germinal enlargement which is not markedly different from the rest of the third ventricle neuroepithelium on day E15 (vm in Fig. 75A). But this neuroepithelial zone becomes conspicuous on day E16 because it remains mitotically more active (Fig. 75B) and is thicker (Fig. 74A) than the dorsally situated neuroepithelium of the diamond-shaped ventricular enlargement. In short-survival thymidine radiograms the ventromedial neuroepithelium remains active on day E17 (Fig. 75C). There is a difference

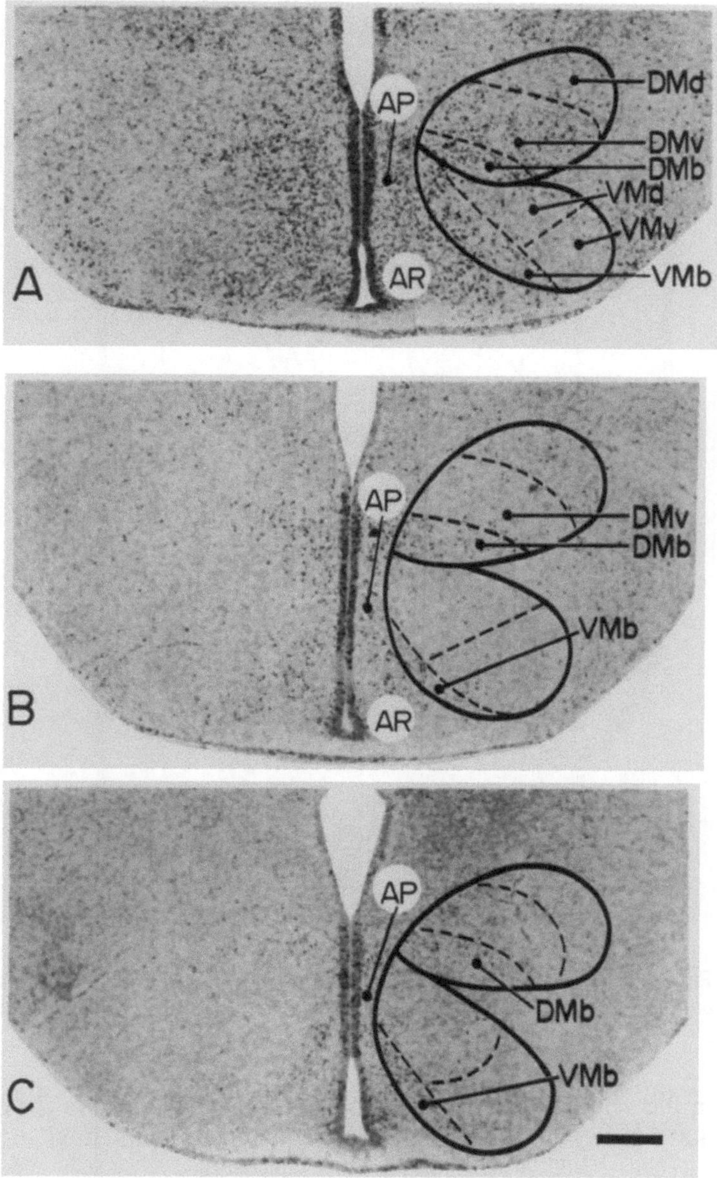

Fig. 72A–C. Coronal thymidine radiograms from perinatal rats labeled on day E15 (**A**), E16 (**B**), and E17 (**C**). The sequential production of neurons and the boundaries of the three components of the dorsomedial nucleus and of the ventromedial nucleus are indicated (*broken outlines*). Paraffin. *Scale*, 200 μm

in the labeling pattern of the ventromedial and dorsomedial neuroepithelium in rats injected on day E16 and killed on day E19 (Fig. 74C) in that the latter retains a large population of labeled cells while in the putative ventromedial neuroepithelium the ependymal cells are unlabeled. In short-survival thymidine radiograms from rats injected on day E18 (Fig. 75D) there are few cells labeled

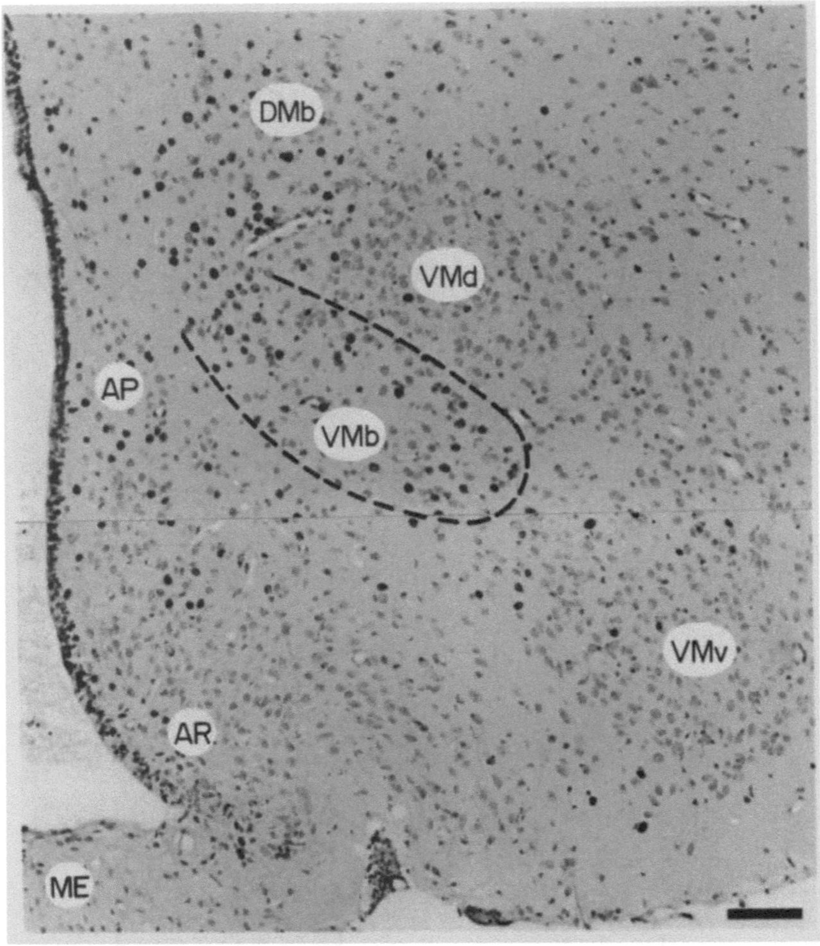

Fig. 73. Delineation of the late-produced basal subnucleus of the ventromedial nucleus (*broken outline*) in an adult rat labeled on days E16+17. Paraffin. *Scale*, 100 μm

in the ventromedial neuroepithelium but many more in the overlying dorsomedial (arrow) and underlying arcuate (ar) neuroepithelium.

Comments. The present observations confirm our earlier quantitative data (Altman and Bayer 1978a: Fig. 7) that the neurons of the ventromedial nucleus are produced between days E13 and E17. We were able to distinguish on the basis of radiographic labeling pattern of their neurons, three subdivisions of the nucleus: the earliest-generated pars ventralis, the intermediate pars dorsalis, and the last-generated pars basalis. The latter region appears to be continuous with the parvocellular periventricular region medially but appears separable from the arcuate nucleus ventrally. The neurons of the ventromedial nucleus originate in the lower half of the neuroepithelial thickening situated between the diamond-shaped ventricle dorsally and the arcuate triangle ventrally. The relation of this neuroepithelial zone to the site of origin of neurons of the dorsomedial nucleus is discussed below.

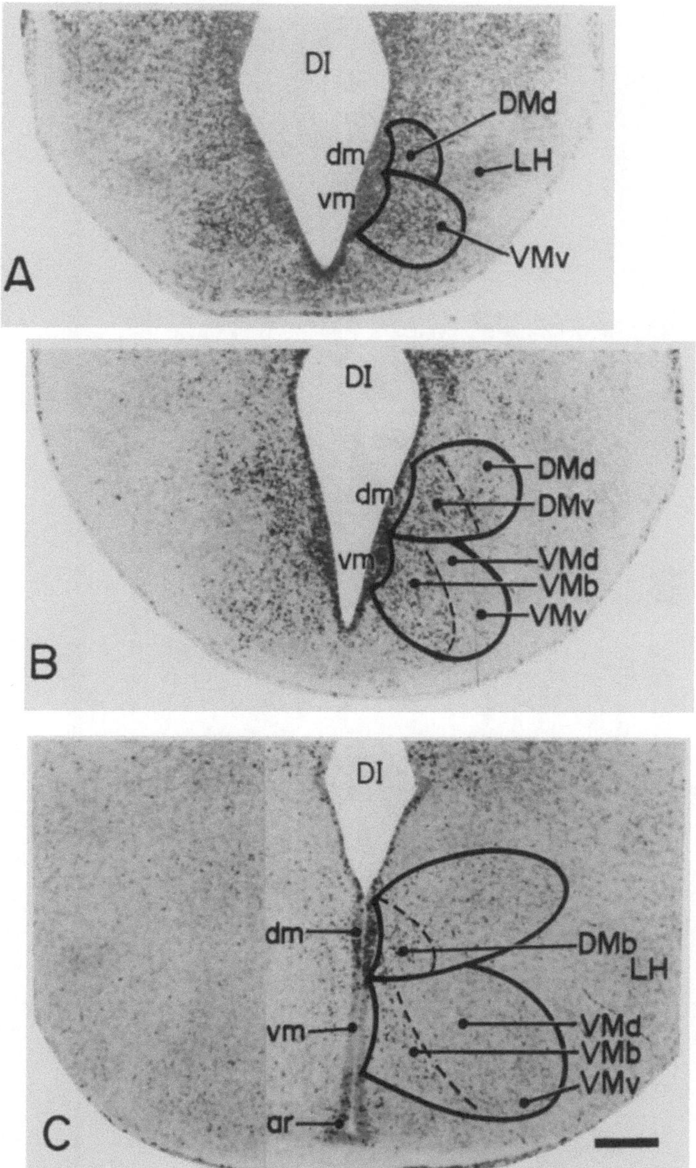

Fig. 74 A–C. Coronal thymidine radiograms from a rat labeled on day E14 and killed on day E16 (**A**), a rat labeled on day E15 and killed on day E17 (**B**), and a rat labeled on day E16 and killed on day E19 (**C**), to show the settling pattern of components of the dorsomedial and ventromedial nuclei. The presumed site of origin of the dorsomedial and ventromedial neurons is also indicated. Paraffin. *Scale*, 200 µm

6.3.2 The Dorsomedial Nucleus

Background. The dorsomedial nucleus has not been studied as extensively as the ventromedial nucleus and there are few indications of its physiological and behavioral functions. Krieg (1932) distinguished a more conspicuous ventral

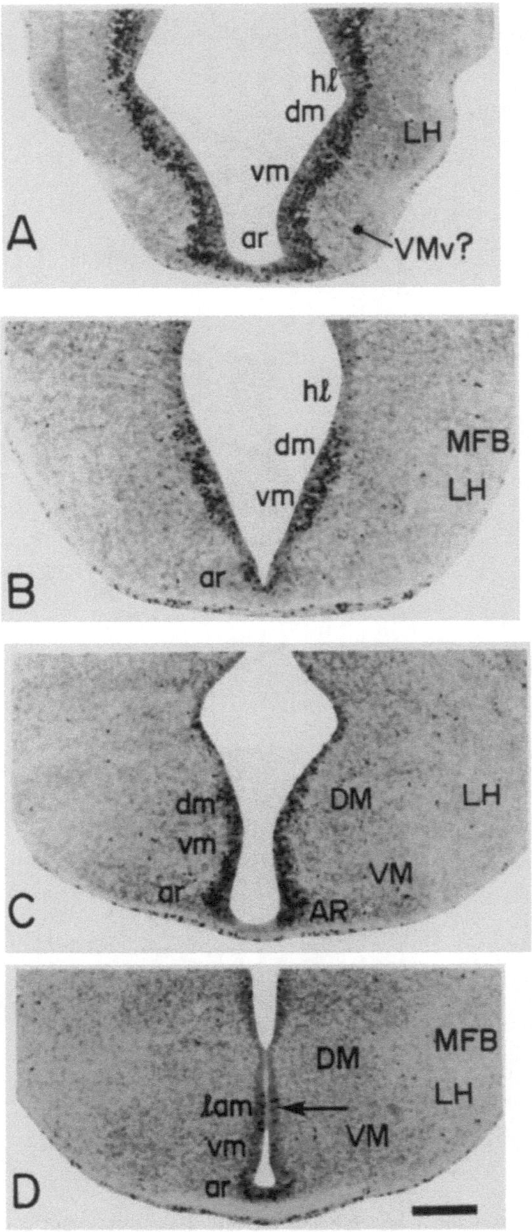

Fig. 75 A–D. Short-survival thymidine radiograms (2-h survival after injection) from rats labeled on day E15 (**A**), E16 (**B**), E17 (**C**) and E18 (**D**) to show the changing configuration of the putative neuroepithelia of the dorsomedial and ventromedial nuclei. *Arrow in* **D** points to the presumed site of origin of the laminated epithelium after cessation of neurogenesis. Paraffin. *Scale*, 200 μm

part of the nucleus and a poorly defined dorsal part. Afferents to the nucleus have been described from a few extrahypothalamic structures, mainly the nucleus accumbens, lateral septal nucleus, bed nucleus of stria terminalis, pontine central gray, and parabrachial area (Kita and Oomura 1982), and from several hypothalamic nuclei, including the ventromedial nucleus (Conrad and Pfaff 1976a, b). Catecholaminergic fibers reach the dorsomedial nucleus from the medulla, pons, and midbrain (Ungerstedt 1971; Hökfelt et al. 1974; Jacobowitz 1975). Its efferents are directed primarily toward other hypothalamic nuclei, including the ventromedial nucleus (Palkovits and Záborszky 1979). Dopamine-containing (Ruggiero et al. 1984) and serotonin-accumulating (Beaudet and Descarries 1979) cell bodies have been described in the medial portion of the dorsomedial nucleus. Ifft (1972) reported a high concentration of heavily labeled cells in the dorsomedial nucleus following injections of [^{3}H]thymidine between days E14 and E17. However, our quantitative studies showed that only a very small percentage of the dorsomedial nuclei are produced on day E14 (Altman and Bayer 1978a: Fig. 7). Both studies agree that there is a dorsolateral-to-ventromedial neurogenetic gradient in this nucleus, one that is a mirror image of that seen in the ventromedial nucleus.

Results. There are few if any heavily labeled cells in the dorsomedial nucleus of fetal rats injected with [^{3}H]thymidine on day E13 (DM in Fig. 71A). More heavily labeled cells are seen in the dorsal part of the nucleus in rats injected on day E14 (DMd in Fig. 71B), but in rats injected on day E15 most of the neurons in the dorsal part of the nucleus are no longer labeled (DMd in Fig. 72A). In rats injected on day E16 the neurons are no longer labeled in the ventral part of the nucleus (DMv in Fig. 72B) except in a narrow basal band (DMb in Fig. 72B). A few cells are still labeled in the latter region in rats injected on day E17 (DMb in Fig. 72C). Accordingly, this strip is designated as the *basal subnucleus of the dorsomedial nucleus*.

In rats injected on day E14 and killed on day E16 (Fig. 74A) there is an aggregate of unlabeled (differentiated) cells in a position corresponding in older embryos to the dorsomedial nucleus. But because the cells of this nucleus are generated after day E14 this cannot be the dorsomedial nucleus (it is designated as the early-produced lateral hypothalamus; LH in Fig. 74A). The first complement of dorsomedial cells leaving the neuroepithelium (dm) are apparently still in the latter's neighborhood (DMd in Fig. 74A) on day E16. An admixture of labeled and unlabeled neurons is present in the dorsal part of the dorsomedial nucleus in rats injected on day E15 and killed on day E17 (DMd in Fig. 74B). In rats injected on day E16 and killed on day E19 only those cells closest to the active neuroepithelial zone are still labeled (dm and DMb in Fig. 74C). The latter region presumably represents the cells of the basal subnucleus that have yet to settle in their final position (compare with rats injected on day E16 or E17 and killed on day E20; Fig. 72B, C).

The neuroepithelium between the diamond-shaped ventricle and the arcuate triangle (DI and ar in Fig. 74) is divisible into two components: an upper half, which could be the source of cells of the dorsomedial nucleus (dm), and a lower half, which might be the source of the ventromedial nucleus (vm). These two are not easy to distinguish from one another in younger embryos injected with [^{3}H]thymidine between days E13 and E15 (Figs. 71, 72A, 74A, B). In some, but not all, sections the two have different labeling patterns in rats injected on day E16 and killed on days E19 (Fig. 74C) or E20 (Fig. 72B), with an indication that the labeled cells are retained longer in the dorsomedial than the ventromedial neuroepithelium.

Comments. The production of neurons of the dorsomedial nucleus begins at an appreciable rate on day E14 and these cells form the dorsal part of the nucleus. The cells generated on day E15 are located mostly in the ventral part of the nucleus, and the last complement of cells, those produced on days E16 and E17, form its narrow basal band. Accordingly, there is similarity in the migratory patterns of the dorsomedial and ventromedial neurons in that in both cases the earliest-generated cells settle farthest laterally and the latest-produced cells form a ventral band closer to the ventricle. But apart from the different direction of cell migration, in one case ventrolaterally and in the other case dorsolaterally, there is a difference in the absolute chronology of neurogenesis in the two nuclei, because the production of dorsomedial neurons begins a day later (E14) than the production of ventromedial neurons (E13; for corroborating quantitative evidence see Altman and Bayer 1978a: Fig. 7). This is relevant to the issue whether or not the neurons of these two nuclei derive from a single neuroepithelial source (Altman and Bayer 1978b: Fig. 10A) or from two separate components of the germinal enlargement situated between the diamond-shaped ventricle and the arcuate triangle.

In some planes of sectioning the neuroepithelium situated in the vicinity of the dorsomedial and ventromedial nuclei appears to be uniform in cell thickness and labeling pattern (Figs. 72A, C, 75A–C). In other regions two components of the neuroepithelium may be distinguished by the following criteria. First, in young embryos the dorsal half of the enlargement is thinner than the ventral half (E16: Fig. 74A; E17: Fig. 74B). Second, in rats injected on day E16 and surviving until day E19 (Fig. 74C) or day E20 (Fig. 72B), the labeled cells are not retained ventrally (suggesting that they are all neurons that migrate away) but are retained dorsally. Finally, in the latter region the ventricular cells can be labeled on day E18 (arrow in Fig. 75D) and for several days thereafter, that is after neurogenesis in this region has ceased. This difference in the two regions can be related to the circumstance that from the germinal zone adjacent to the dorsomedial nucleus a specialized ventricular structure develops, the laminated epithelium (Sect. 8). A possible interpretation of these observations is that there are two cell lines in the germinal enlargement considered. The dorsomedial cell line first produces dorsomedial neurons and then the laminated epithelium, while the ventromedial cell line first produces ventromedial neurons and then a specialized, presumably tanycytic epithelium (Altman and Bayer 1978c Figs. 2, 6; Sect. 8). But the fact that the difference in the labeling pattern of the regions is seen only at restricted levels and in some planes of sectioning indicates that the neuroepithelium extending between the diamond-shaped ventricle and the arcuate triangle has many other components in addition to the cell lines of the dorsomedial and ventromedial nuclei. We have identified earlier as an anterior derivative of this germinal region the anterior periventricular area (Figs. 58, 59).

6.4 Development of the Posterior Hypothalamus: Ventral Region

The components of the ventral region of the posterior hypothalamus derive from its corresponding neuroepithelium (phv in Figs. 3, 4, 7, 8), which is clearly demarcated in the midline third ventricle by the infundibular recess (ir) anteriorly and the mammillary recess (mr) and mammillary body neuroepithelium (ma) posterodorsally (Figs. 3A, B, 7A, B, 8A). The components of this region are the subthalamic nucleus, the tuberal magnocellular nucleus, and the premammillary nuclei.

6.4.1 The Subthalamic Nucleus

Background. The term subthalamus was introduced by Forel (1877) as an inclusive designation for the zona incerta, fields H_1 and H_2 (of Forel), and the corpus Luysi (nucleus of Luys). Herrick (1910) referred to this region situated between the dorsal thalamus and the hypothalamus as the ventral thalamus, and this term has gained wide currency. However, the nucleus of Luys has become generally known as the subthalamic nucleus. As an unfortunate new complication, we propose to include the subthalamic nucleus with the hypothalamus (instead of the subthalamus or ventral thalamus) because of the evidence to be presented below that its neurons derive from the posterodorsal hypothalamic neuroepithelium. This is not an altogether novel classification, since some developmentally oriented anatomists have suspected such a derivation. Thus Rose (1942), who called this general region the dorsal hypothalamic area, entertained the possibility that the subthalamic nucleus derives from the hypothalamus proper (Rose 1942 p. 104), and Kuhlenbeck (1977 pp. 301–302) lists the nucleus of Luys with the "dorsal (hypothalamic) and entopeduncular complex."

The rat subthalamic nucleus is a compact, lens-shaped structure, apposed to the medial wall of the cerebral peduncle beneath the zona incerta. It consists of two types of densely packed cells (Kita et al. 1983a). One cell type has dendrites extending into the cerebral peduncle, and bifurcating axons that reach the globus pallidus and the entopeduncular nucleus rostrally, and the substantia nigra caudally. The axons of the other cell type arborize within the subthalamic nucleus. The major afferent supply of the subthalamic nucleus, in monkey and cat, is from the globus pallidus (Whittier and Mettler 1949a; Grofová 1969), the substantia nigra (Afifi and Kaelber 1965; Rinvik et al. 1979), and the motor cortex and adjacent frontal areas (De Vito and Smith 1964; Petras 1967; Hartmann-von Monakow et al. 1978; Künzle 1978; Carpenter et al. 1981). Recent studies in the rat have confirmed afferent projection to the subthalamic nucleus from the globus pallidus (Carter and Fibiger 1978; Kita et al. 1983b) and the motor cortex (Kitai and Deniau 1981; Afsharpour 1985). Efferents from the subthalamic nucleus have been traced in monkey and cat to the globus pallidus and the substantia nigra (Nauta and Cole 1978; Carpenter et al. 1981). These projections have been confirmed in the rat with electrophysiological (Deniau et al. 1978; Perkins and Stone 1980; Kita et al. 1983b) and anatomical (van der Kooy and Hattori 1980) techniques. Clinical (Martin 1927) and experimental (Whittier and Mettler 1949b; Carpenter et al. 1950) studies have implicated the subthalamic nucleus in dyskinesias (hemiballism).

Results. In rats that received [^{3}H]thymidine on day E15 and were killed on day E22 most of the cells of the subthalamic nucleus were no longer labeled rostrally (Figs. 76A, 77A) while the majority were labeled caudally (Figs. 76D, 77D). The few labeled cells seen rostrally were concentrated in the dorsal and

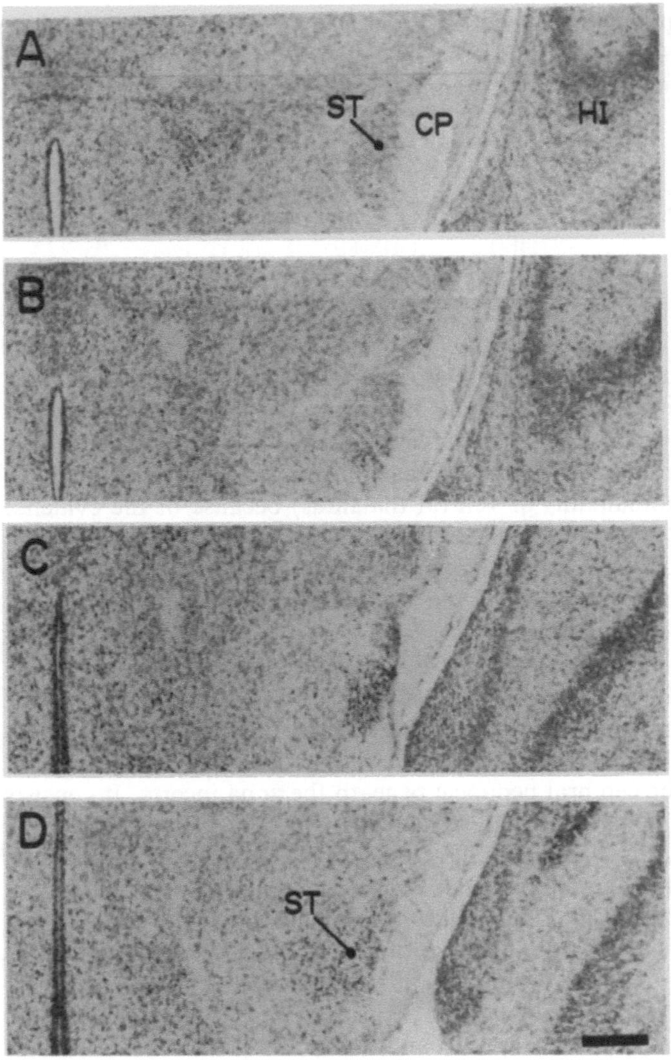

Fig. 76 A–D. Coronal thymidine radiograms, from rostral (**A**) to caudal (**D**), from a perinatal rat labeled on day E15. The subthalamic neurons are no longer labeled rostrally (**A**) except in its dorsal and ventral tips. In the midportion of the nucleus (**B, C**) the ventromedial neurons are labeled, and caudally (**D**) most of the subthalamic neurons are labeled. Paraffin. *Scale*, 200 μm

ventral tips of the nucleus (Fig. 77A). In the mid-portion of the subthalamic nucleus (Figs. 76B, C, 77B, C), the unlabeled cells were located dorsally and the labeled cells ventrally. This indicates a double neurogenetic gradient within the nucleus from rostral and dorsolateral (early neurons) to caudal and ventromedial (late neurons). An exception to this is the labeled dorsally situated cells at rostral levels; this region is designated as the *accessory subthalamic nucleus* (STa in Fig. 77A, B).

In rats that received [^{3}H]thymidine on day E15 and were killed on day E16, some unlabeled subthalamic neurons (those that were generated on day

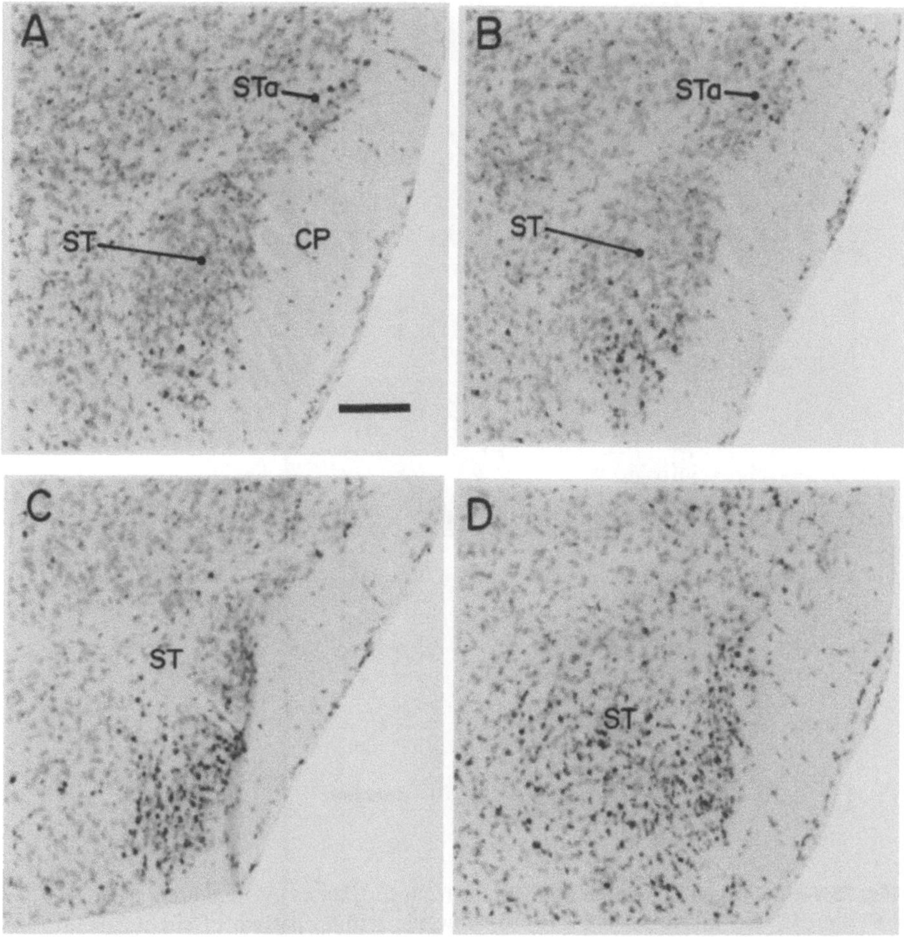

Fig. 77 A–D. Detail of Fig. 76 to show the neurogenetic gradient in the subthalamic nucleus and the accessory subthalamic nucleus. Paraffin. *Scale*, 100 μm

E14 or earlier) have already settled rostrally (STr in Fig. 78 A) and at intermediate levels (STi in Fig. 78 B), but caudally the future site of the subthalamic nucleus is still devoid of cells (broken outline in Fig. 78 C). In rats injected on day E15 and killed on day E17 there is an apparent increase in the cell population of the subthalamic nucleus with an admixture of unlabeled and labeled cells at anterior intermediate levels (Fig. 79 A) and a preponderance of labeled cells more caudally (Fig. 79 B). At most caudal levels cells are apparently still migrating (arrow in Fig. 79 C). The subthalamic neurons appear to originate in a portion of the neuroepithelium in the anterodorsal region of the mammillary recess (st in Fig. 79 C). This site is caudal to the site of origin of the premammillary nuclei (pm in Fig. 79 B) and rostral to the site of origin of the cells of the tuberal magnocellular nucleus (Figs. 86, 87). The spindle-shaped migratory cells near their site of origin caudally are shown in a methacrylate-embedded radiogram in Fig. 80. The curved, caudal-to-rostral course of these migrating cells is illustrated in Figs. 81 and 82.

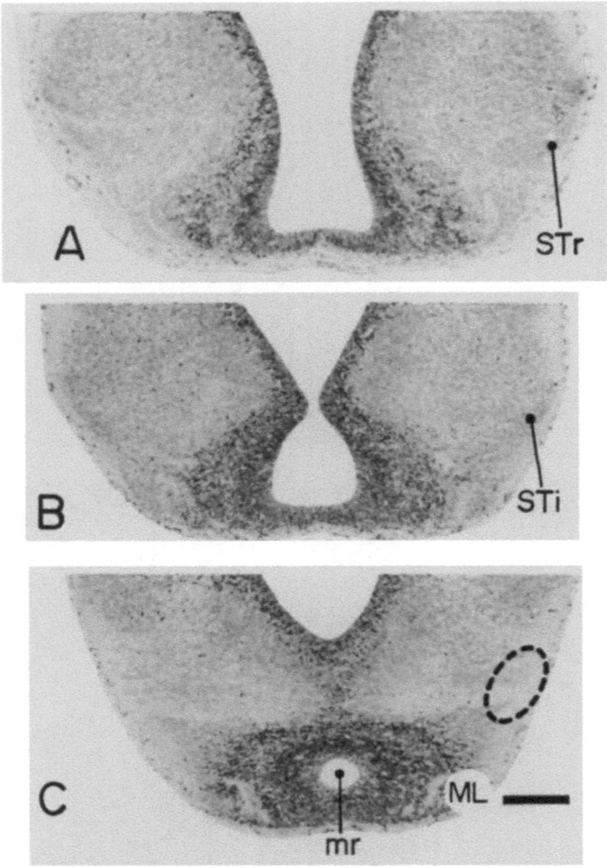

Fig. 78A–C. Coronal thymidine radiograms, from rostral (**A**) to caudal (**C**), from a rat tagged on day E15 and killed on day E16, to show the settling pattern of the unlabeled (earlier-produced) subthalamic neurons. Broken outline (**C**) future site of the caudal part of the subthalamic nucleus. Paraffin. *Scale*, 200 μm

In rats injected on day E16 and killed on day E22 the subthalamic neurons are not labeled rostrally either in the main nucleus or the accessory nucleus (Fig. 83A) but a few neurons are still labeled in the most caudal tip of the nucleus ventrally (Fig. 83B).

Comments. These observations indicate that the neurons of the subthalamic nucleus derive from a neuroepithelial zone in the anterodorsal region of the mammillary recess, caudal to the site of origin of the neurons of the premammillary nuclei and rostral to the site of origin of the neurons of the tuberal magnocellular nucleus. Since the mammillary recess is undisputably part of the hypothalamic rudiment, and since the other structures originating at this site, the premammillary nuclei and the tuberal magnocellular nucleus, are part of the hypothalamus, the conclusion follows that, ontogenetically, the subthalamic nucleus should be considered part of the "hypothalamus proper."

The traditional inclusion of the subthalamic nucleus with the ventral thalamus (or subthalamus) is understandable in view of its location in the vicinity

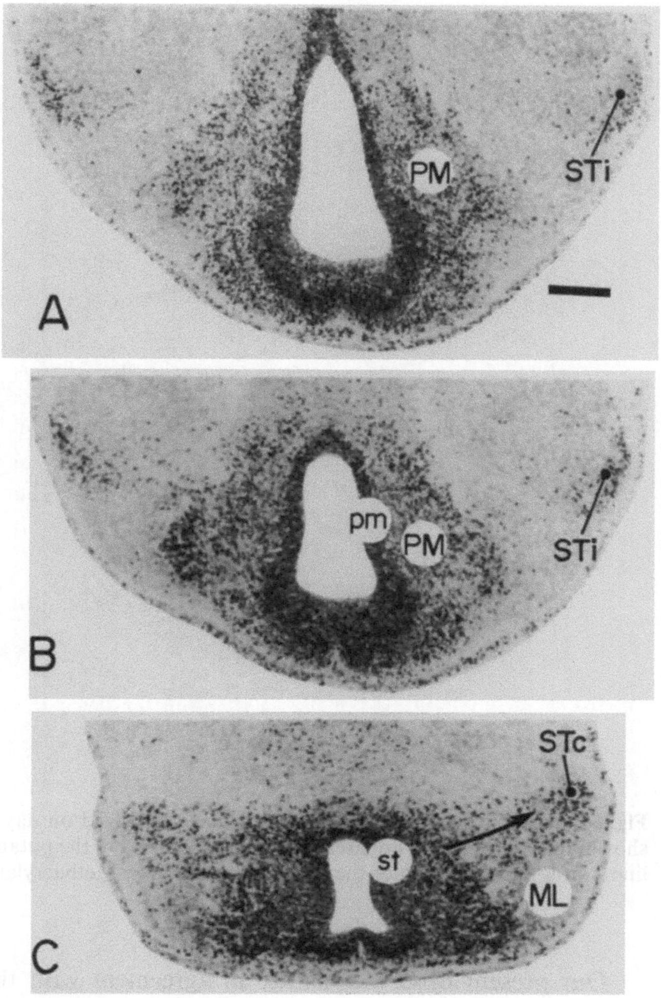

Fig. 79 A–C. Coronal thymidine radiograms, from rostral (**A**) to caudal (**C**), from a rat tagged on day E15 and killed on day E17, to show the settling pattern of labeled (later-produced) subthalamic neurons and their origin caudally (*arrow in* **C**) from the anterodorsal region of the mammillary recess. Paraffin. *Scale*, 200 μm

of the zona incerta. Another consideration may have been a functional one, namely, that the subthalamic nucleus, like other components of the traditional ventral thalamus, is implicated in somatic motor functions and not in visceral and humoral functions, as are so many hypothalamic nuclei. Such a functional consideration was the basis of the classification of the subthalamic nucleus by some investigators with the entopeduncular nucleus (Kuhlenbeck 1977) and even with the globus pallidus (Spatz 1925; Richter 1965). Although this functional link is reinforced by the reciprocal connections of the subthalamic nucleus with the globus pallidus, entopeduncular nucleus, and substantia nigra (reviewed earlier), these three "extrapyramidal" structures lying along the corticospinal tract (internal capsule and cerebral peduncle) derive from three distinct primordia (telencephalic, diencephalic, and mesencephalic), not from a single neuroepithelial source.

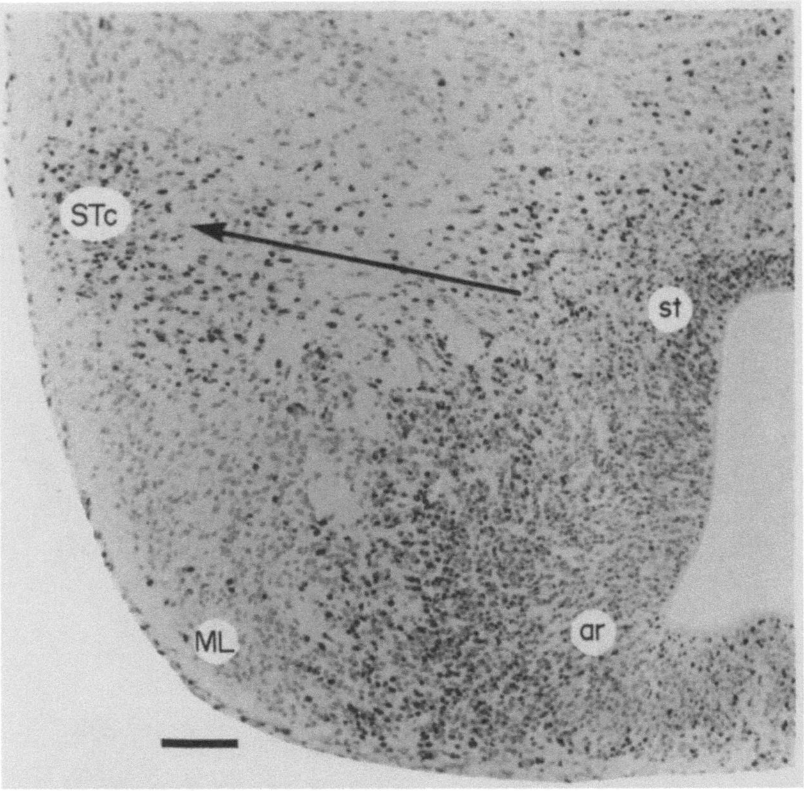

Fig. 80. Coronal thymidine radiogram from a rat tagged on day E15 and killed on day E17, showing spindle-shaped, apparently migrating cells from the putative subthalamic neuroepithelium to the caudal region of the subthalamic nucleus. Methacrylate. *Scale*, 100 μm

Our present observations are in agreement with the previous quantitative conclusion that the bulk of subthalamic neurons are produced on days E14 and E15 (Altman and Bayer 1979a: Fig. 10), and they extend the earlier finding that there is a neurogenetic gradient within the nucleus from dorsolateral to ventromedial (Altman and Bayer 1979b: Figs. 10, 11). The postmitotic subthalamic neurons apparently migrate by a semicircular route from the anterodorsal mammillary recess neuroepithelium posteroventrally to their final location (Figs. 79–83) rostrally and dorsally. The cells settle in a modified "outside-in" pattern, the earliest-generated neurons becoming situated most rostrally and dorsolaterally and the later-generated neurons caudally and ventromedially. The first complement of cells, those produced on day E14 (or earlier), appear to have settled by day E16 (Fig. 78A, B), and those produced on day E15 by day E17 (Fig. 79A, B), indicating a minimum migratory time of about 2 days. Since the corticospinal tract fibers do not reach the level of the cerebral peduncle for many more days (they are present by day E22; Figs. 76, 77), a guiding influence by these fibers must be ruled out.

Several recent studies have shown that there is a topographical organization in the subthalamic nucleus in terms of both its efferent projections and afferent supply. Thus, Carpenter et al. (1981) reported that, in the monkey, subthalamic

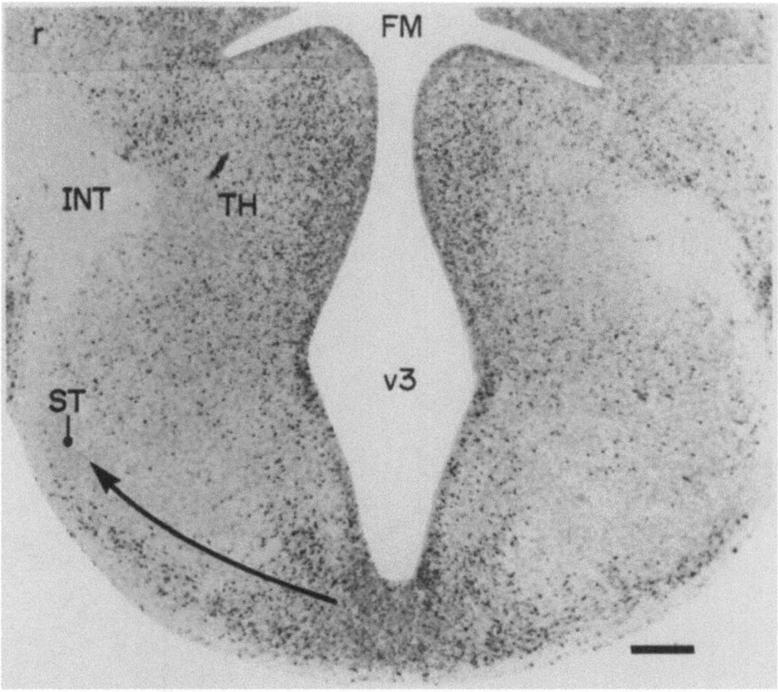

Fig. 81. Horizontal thymidine radiogram, from a rat labeled on day E15 and killed on day E17, showing the curved, caudal-to-rostral course (*arrow*) of the migrating subthalamic neurons. Methacrylate. *Scale*, 200 μm

efferents to the globus pallidus originate mainly from neurons situated in the medial two-thirds of the nucleus and that pallidal afferents terminate in distinctive loci of the subthalamic nucleus. In the rat, van der Kooy and Hattori (1980) observed that the projection from the lateral subthalamic nucleus is to the lateral pallidum and they suggested, by implication, that the projection from the medial subthalamic nucleus is to the medial globus pallidus. Hartmann-von Monakow et al. (1978) reported a topographic organization of fibers from different regions of the monkey motor cortex to the subthalamic nucleus. A topographical organization of cortical fibers distributing to the subthalamic nucleus has also been reported in the rat (Afsharpour 1985). The exact relationship of these afferent projection patterns in relation to the early- and late-produced neurons of the subthalamic nucleus is at present unclear. With regard to efferent outflow, van der Kooy and Hattori (1980) found in the rat that medial subthalamic neurons project mainly to the medial part of the pars compacta of the substantia nigra. Insofar as the neurogenetic gradient in the substantia nigra (Altman and Bayer 1981: Fig. 30) is very similar to that of the subthalamic nucleus, in this case the pattern might be that the early-produced subthalamic neurons establish connections with the early-produced nigral neurons and vice versa. The possibility that the early- and late-generated subthalamic neurons represent subpopulations with different topographical connections deserves to be investigated.

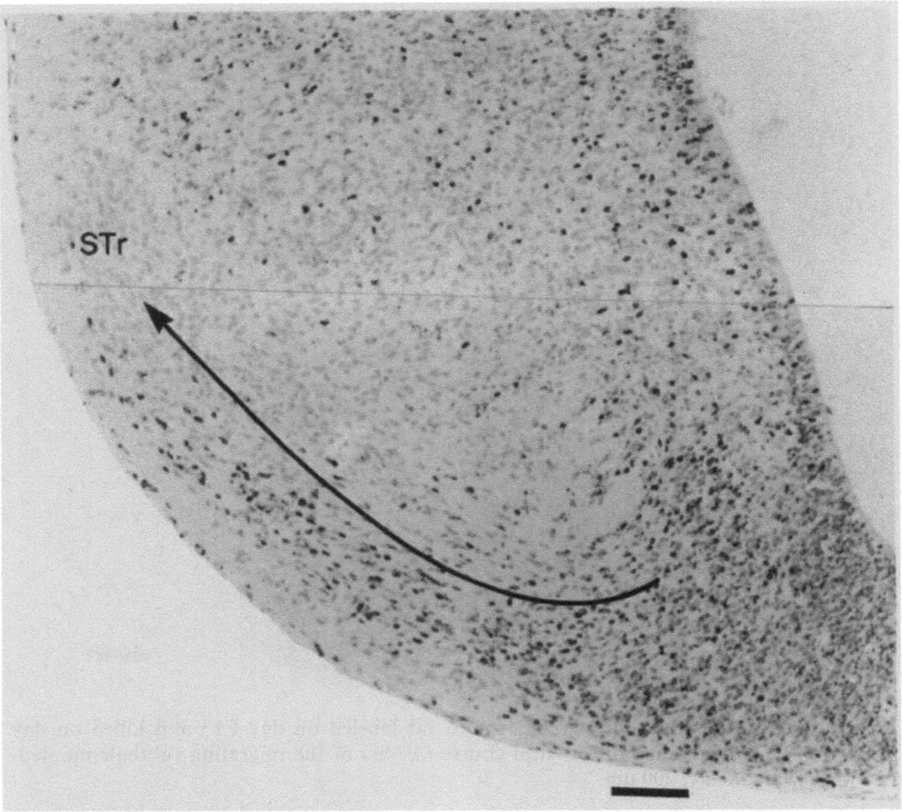

Fig. 82. Detail of Fig. 81. *Scale*, 100 μm

6.4.2 The Tuberal Magnocellular Nucleus

Background. The tuberal magnocellular nucleus is situated rostral to the lateral mammillary nucleus ventrolaterally and is composed of cells that stain intensely but are somewhat smaller than the magnocellular neurons of the supraoptic and paraventricular nuclei. This nucleus has yet to receive the attention of investigators. Christ (1969) has referred to it as the tuberomammillary nucleus, a term that we have adopted in our previous developmental study (Altman and Bayer 1978a). Bleier et al. (1979) refer to it as the caudal magnocellular nucleus and distinguish a subcomponent as the postmammillary caudal magnocellular nucleus. Naik (1975) describes a group of large neurons in the "ventrolateral-premammillary body" as luteinizing-hormone releasing-hormone positive cells. However, a perusal of available studies of retrograde transport of tracers from the median eminence (Wiegand and Price 1980; Lechan et al. 1982) seems to rule out the possibility that these cells, resembling neurosecretory elements, convey releasing hormones to the anterior pituitary. Our quantitative studies have shown that the neurons of this nucleus (Altman and Bayer 1978a: Fig. 8; tuberomammillary nucleus) are generated between days E15 and E18, with peaks on days E16 and E17.

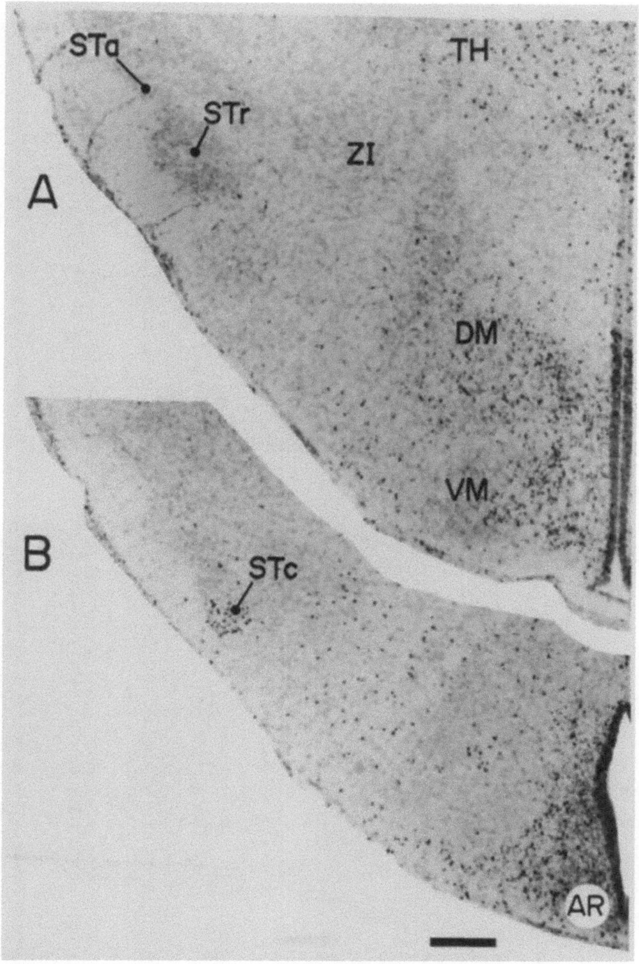

Fig. 83 A, B. Coronal thymidine radiograms, from rostral (**A**) to caudal (**B**), from a perinatal rat tagged on day E16, showing the caudal location of the latest-generated neurons of the subthalamic nucleus. Paraffin. *Scale*, 200 µm

Results. The position of the tuberal magnocellular nucleus (TU) is illustrated in coronal sections rostrally (Fig. 84A) and caudally (Fig. 84B) in an adult rat that received [^{3}H]thymidine on days E16+17. In this group the majority of its neurons are heavily or lightly labeled. The position of the nucleus in relation to the lateral mammillary nucleus (ML) is illustrated in a horizontal section from an adult rat that was labeled on days E17+18 (Fig. 85A). Many of the tuberal magnocellular neurons are still labeled (Fig. 85B). With injections made after day E18, these neurons are no longer labeled (not shown). The site of origin of the tuberal magnocellular neurons in the dorsal neuroepithelium of the mammillary recess, and a straight mediolateral migratory route, is suggested by the radiograms from a rat injected on day E16 and killed on day E18 (Fig. 86). The spindle-shaped cells appear to migrate as far as the ventral tip of the subthalamic nucleus (ST in Fig. 87). The cells generated on day E16, and those earlier formed, constitute the unlabeled elements of the tuberal mag-

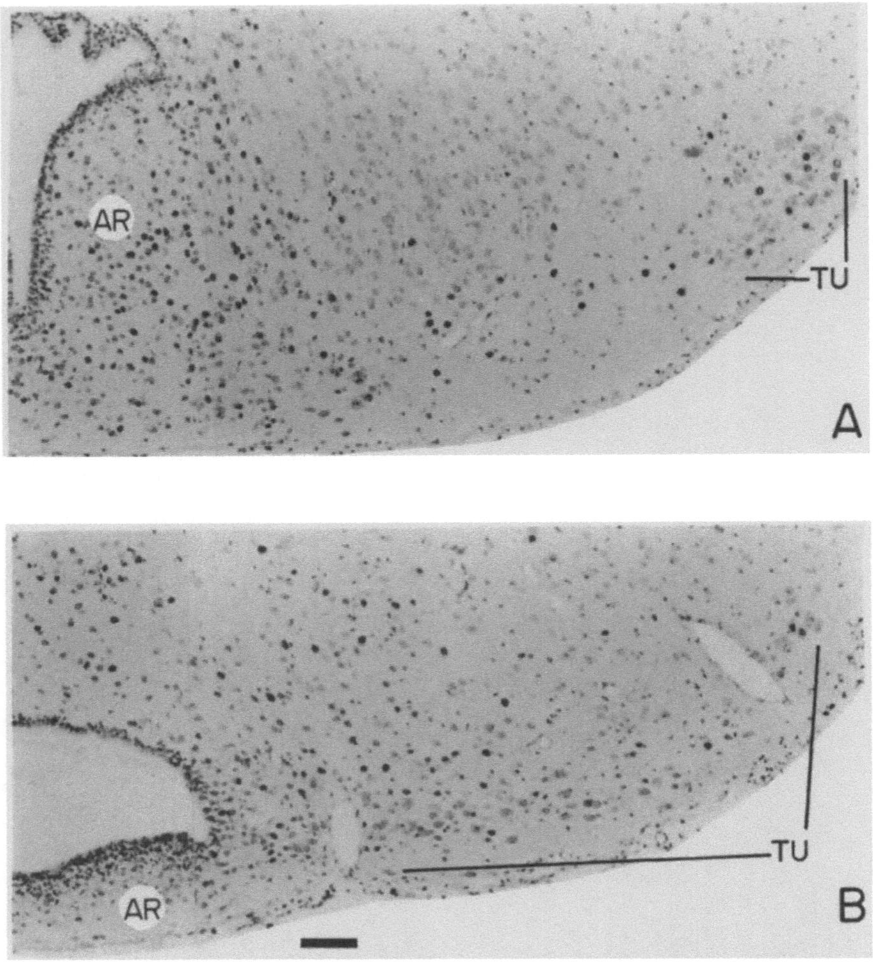

Fig. 84 A, B. Coronal thymidine radiograms, from rostral (**A**) to caudal (**B**), from an adult rat that was labeled on days E16 + 17. Note the far-lateral position of the tuberal magnocellular neurons rostrally (**A**) and their relation to the arcuate nucleus caudally (**B**). Paraffin. *Scale*, 100 μm

nocellular nucleus, as seen in the rat injected on day E17 and killed on day E18 (Fig. 88 A). In rats injected on day E17 and killed on day E20 (Fig. 88 B) labeled cells mingle with unlabeled cells in the tuberal magnocellular nucleus, indicating an absence of a gradient in the settling of these neurons.

Comments. The present results confirm the late production of the tuberal magnocellular neurons. In this respect these neurons, which show some of the cytological features of the magnocellular cells of the supraoptic and paraventricular nuclei, resemble more the late-produced parvocellular cells of the arcuate nucleus (Sect. 7.2). While we have not been able to precisely delineate the site of origin of these large cells, they originate from the caudal tip of the mammillary recess, dorsal to the site of production of the arcuate neurons. Our observations suggest that these magnocellular elements migrate rapidly over a straight route to their ventrolateral settling position.

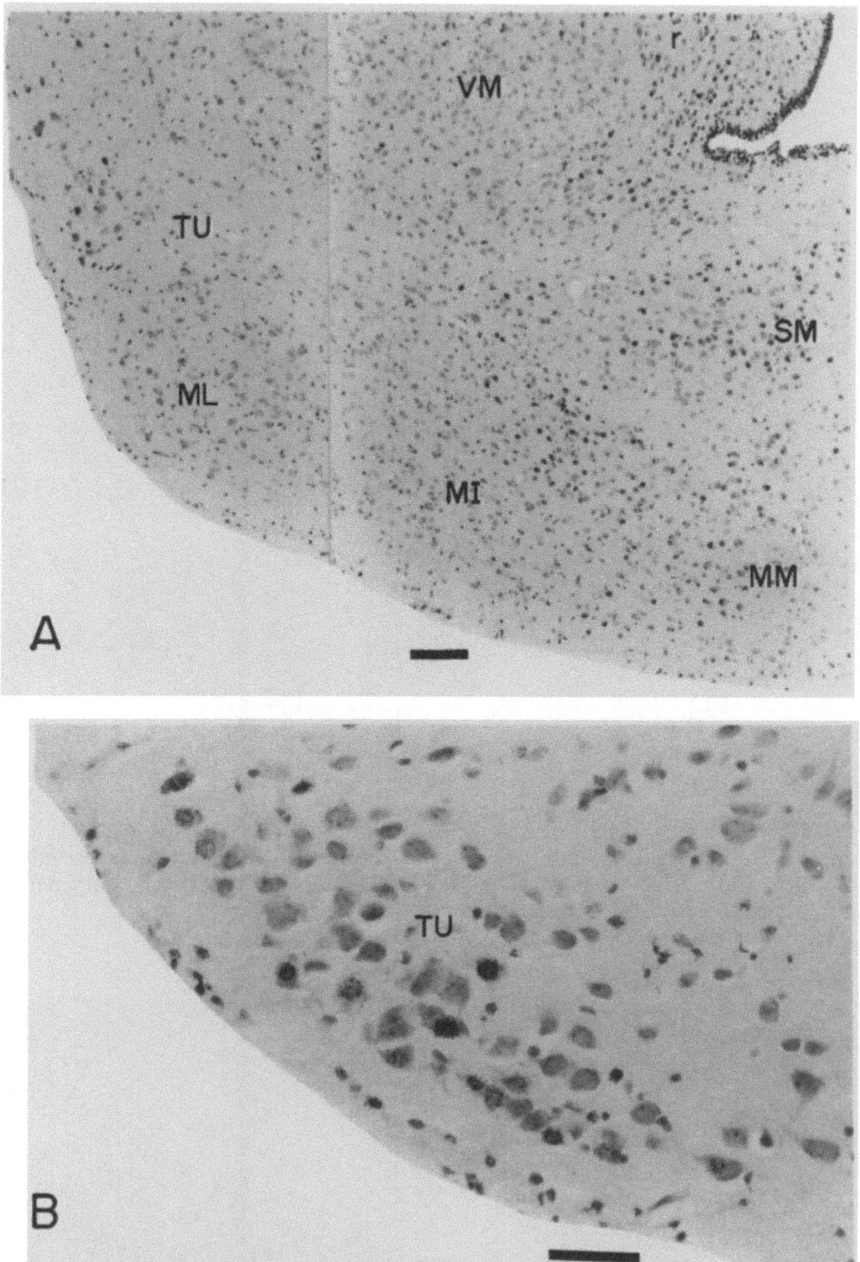

Fig. 85. A Horizontal thymidine radiogram, from an adult rat tagged on days E17+18. Note the position of the tuberal magnocellular nucleus in relation to the lateral mammillary nucleus. **B** Detail. A high proportion of the tuberal magnocellular neurons are still labeled. Paraffin. *Scales*, **A** 100 μm; **B** 50 μm

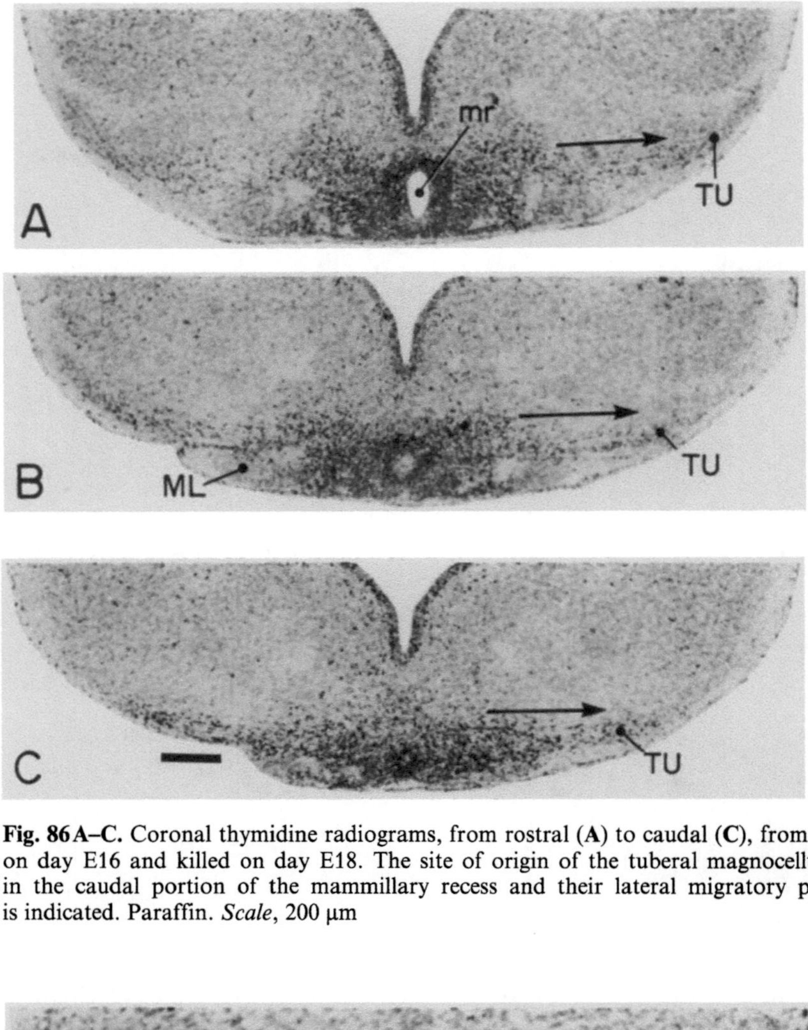

Fig. 86 A–C. Coronal thymidine radiograms, from rostral (**A**) to caudal (**C**), from a rat tagged on day E16 and killed on day E18. The site of origin of the tuberal magnocellular neurons in the caudal portion of the mammillary recess and their lateral migratory path (*arrows*) is indicated. Paraffin. *Scale*, 200 μm

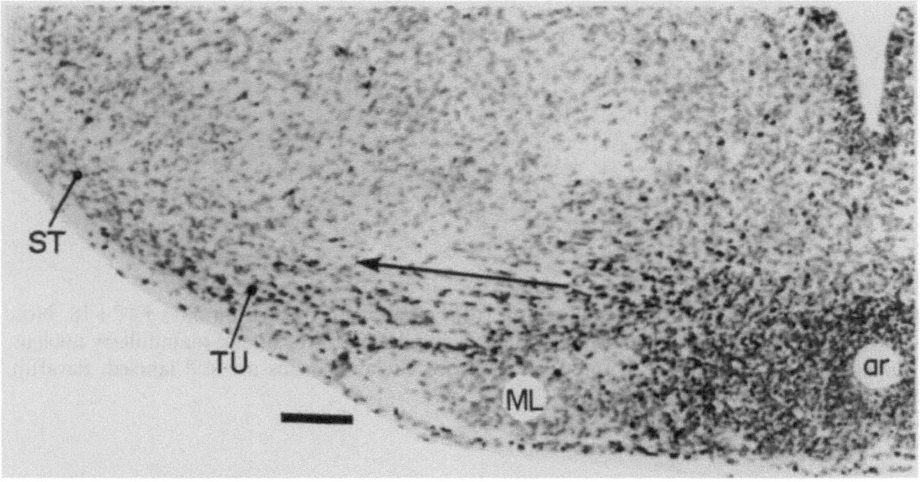

Fig. 87. The migratory path of tuberal magnocellular neurons (*arrow*) in a rat tagged on day E16 and killed on day E18 at higher magnification. The contiguity with the earlier-produced neurons of the subthalamic nucleus is also indicated. Paraffin. *Scale*, 100 μm

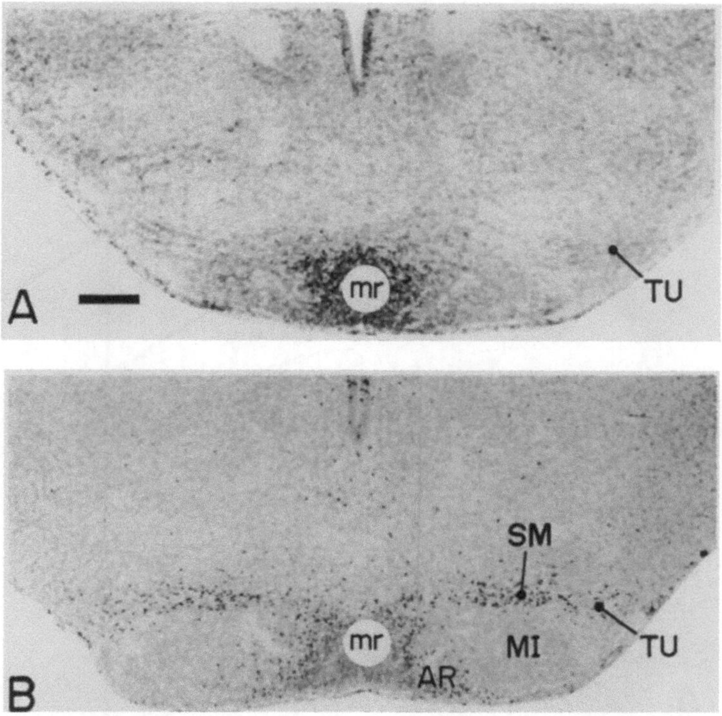

Fig. 88 A, B. Coronal thymidine radiograms from rats tagged on day E17 and killed on day E18 (**A**) and day E20 (**B**). The relatively slow migratory speed of the tuberal magnocellular neurons, and an absence of a neurogenetic gradient, is indicated. Paraffin. *Scale*, 200 μm

6.4.3 The Premammillary Nuclei

Background. Gurdjian (1927) distinguished in the rat between a dorsal and a ventral premammillary nucleus. He noted in Golgi material that axons of the ventral premammillary nucleus extend to the periventricular region. In a more recent Golgi study, Millhouse (1979) described axons from premammillary neurons to the adjacent hypothalamic nuclei, including the arcuate nucleus. In tracer studies (reviewed by Palkovits and Záborszky 1979) afferents were noted that reach the premammillary nuclei from several hypothalamic regions, including the arcuate nucleus, the hippocampus, the septum, the interstitial nucleus of the stria terminalis, the amygdala, and the pontine tegmentum. The efferents of the premammillary nuclei were traced to the supraoptic, anterior hypothalamic, ventromedial, and arcuate nuclei. Our quantitative radiographic studies in adult rats (Altman and Bayer 1978a: Fig. 8) indicated that the neurons of the ventral and dorsal premammillary nuclei are generated between days E13 and E15, neuron production being somewhat earlier in the ventral nucleus than in the dorsal nucleus.

Results. The two components of the premammillary nuclei are easily distinguished in thymidine radiograms from rats injected between days E15 and E17 and killed on day E22 (Figs. 89, 90) by their different labeling patterns. The

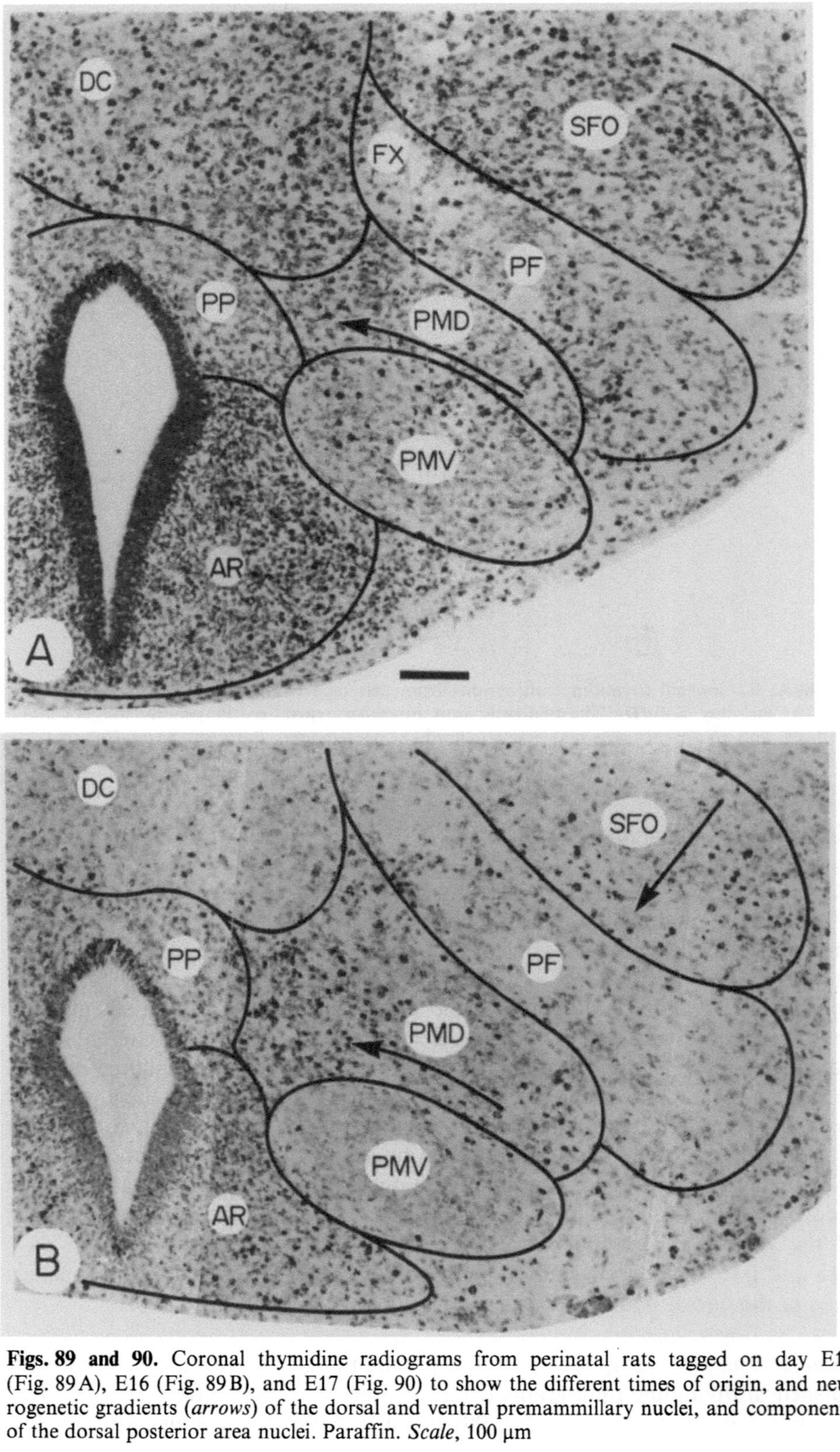

Figs. 89 and 90. Coronal thymidine radiograms from perinatal rats tagged on day E15 (Fig. 89A), E16 (Fig. 89B), and E17 (Fig. 90) to show the different times of origin, and neurogenetic gradients (*arrows*) of the dorsal and ventral premammillary nuclei, and components of the dorsal posterior area nuclei. Paraffin. *Scale*, 100 µm

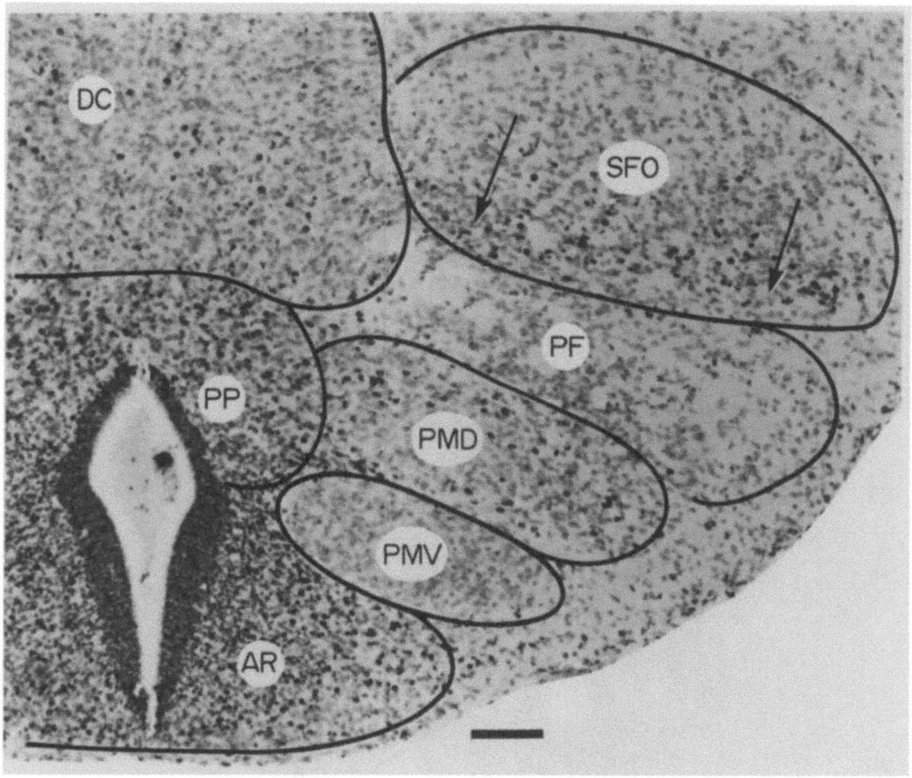

Fig. 90. Legend see Fig. 89

neurons of both the dorsal nucleus and the ventral nucleus, as well as those of adjacent structures, are all labeled in rats injected on day E14 (not shown). In perinatal rats injected on day E15 most of the neurons of the ventral nucleus (PMV in Fig. 89A) are no longer labeled and the few that are show heavy labeling. In contrast, in the dorsal nucleus (PMD in Fig. 89A) all cells are labeled, those situated laterally somewhat more heavily than those situated medially. In perinatal rats injected on day E16 (Fig. 89B) only an occasional cell is labeled in the ventral nucleus but many heavily labeled cells are seen in the dorsal nucleus. A decreasing proportion of the dorsal nucleus neurons is still labeled in rats injected on day E17 (Fig. 90) but no longer in rats injected on day E18 (not shown). These observations establish that the neurons of the ventral premammillary nucleus are generated predominantly on day E14 with a smaller proportion acquired on day E15, whereas the neurons of the dorsal premammillary nucleus are produced mostly on days E15 and E16 with a fair proportion added on day E17. In the available material we could not discern a clear neurogenetic gradient within the ventral nucleus. In the dorsal nucleus a lateral-to-medial gradient is indicated (arrows in Figs. 89A, B) but there appears to be present an aggregate of scattered labeled cells that does not follow this gradient (Figs. 89B, 90).

The site of origin of neurons of the premammillary nuclei (and of some other nuclei to be described in the next section) appears to be a neuroepithelial zone dorsal to the germinal source of arcuate neurons (ar) and rostral to the

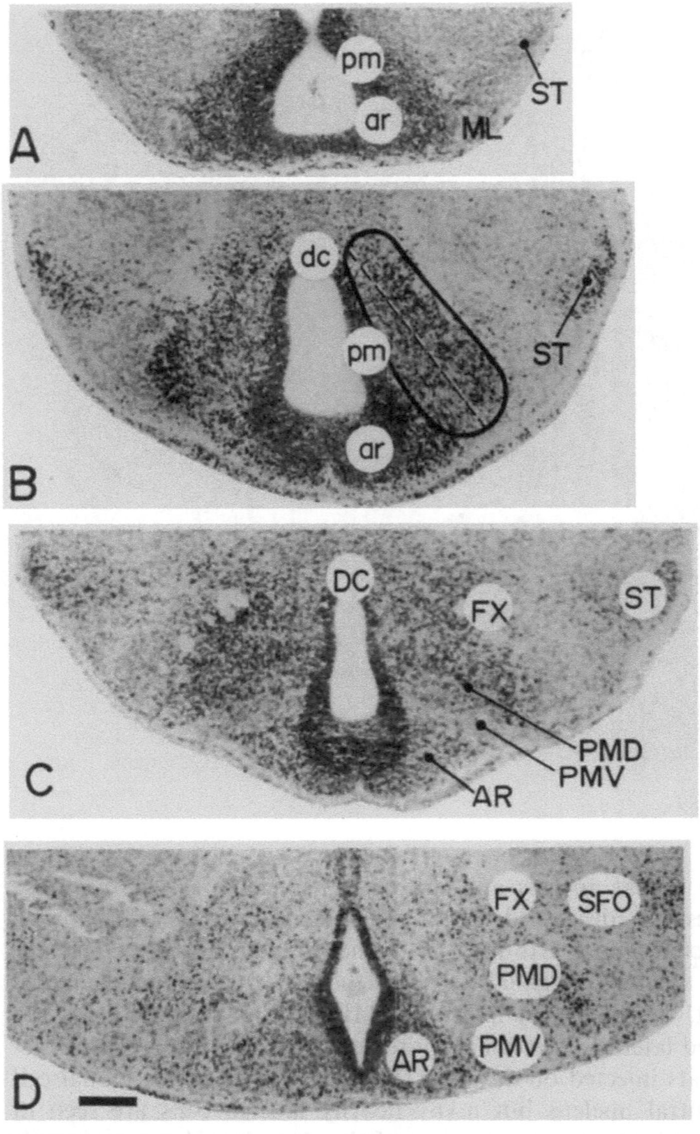

Fig. 91 A–D. Coronal thymidine radiograms from rats tagged on day E15 and killed on day E16 (**A**), E17 (**B**), E18 (**C**), and E20 (**D**). The outflow of differentiating neurons (with two distinguishable zones) from the presumptive premammillary neuroepithelium begins on day E17 (**B**), and by day E18 (**C**) the earlier-generated (unlabeled) neurons of the ventral premammillary nucleus and the later-generated (labeled) neurons of the dorsal premammillary nucleus are beginning to settle in their final location. Paraffin. *Scale*, 200 μm

mammillary neuroepithelium (Figs. 91, 92). In rats injected on day E15 and killed on day E16 (Fig. 91 A), the thick germinal matrix of the putative premammillary neuroepithelium (pm) is not markedly different from the underlying arcuate neuroepithelium and the differentiating zone around it is not yet conspicuous. In rats injected on day E15 and killed on day E17 (Fig. 91 B) the thickness of the premammillary neuroepithelium is greatly reduced with respect to the arcuate neuroepithelium. The adjacent differentiating zone has two components,

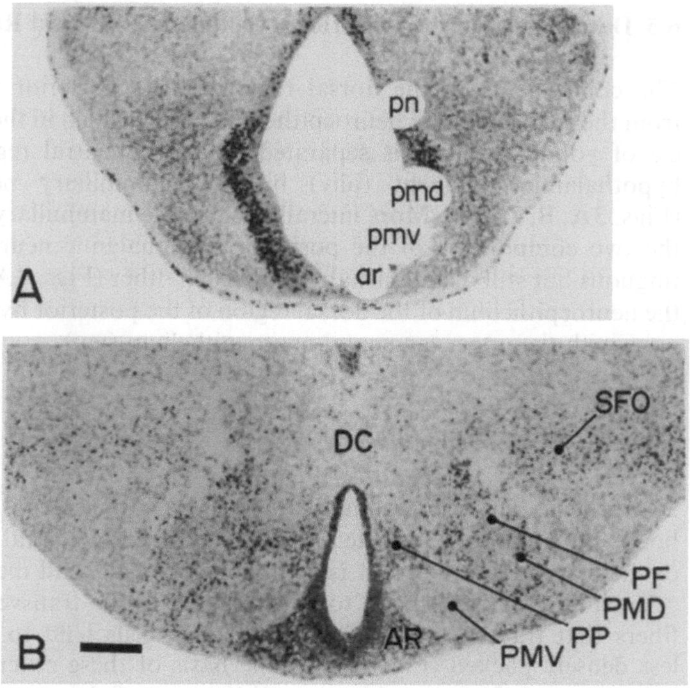

Fig. 92 A, B. Coronal thymidine radiograms from rats tagged on day E16 and killed 2 h later (**A**) and on day E20 (**B**) to show the possible segregation of the neuroepithelial zones of the posterior area nuclei, the premammillary nuclei, and the arcuate nucleus. Paraffin. *Scale*, 200 µm

an outer band of mostly heavily labeled cells and an inner band with an admixture of unlabeled and heavily labeled cells. In rats injected on day E15 and killed on day E18 (Fig. 91 C), many of the migratory cells apparently have reached their targets and the dorsal and ventral premammillary nuclei can be distinguished by their differential labeling patterns. The pattern seen in rats injected on day E15 and killed on day E20 (Fig. 91 D) resembles the perinatal pattern (Fig. 89 A). The possible segregation of the premammillary neuroepithelium into a ventral (pmv) and a dorsal (pmd) component is suggested by observations in rats injected on day E16 and killed 2 h later (Fig. 92 A). The dorsal premammillary nucleus, by virtue of its presumed site of origin and later production of its cells, may be related to the neurons of the posterior periventricular area (PP in Fig. 92 B).

Comments. The present results extend and modify our earlier quantitative results in adult rats (Altman and Bayer 1978: Fig. 8). The difference in the time of origin of the ventral and dorsal premammillary nuclei seems more pronounced than the earlier results indicated. This discrepancy may be due to the circumstance that the boundaries of the two nuclei are less obvious in adults than in late fetuses and we may have failed adequately to distinguish the two. The nature of the interrelation of these two contiguous nuclei, which are presently related only by a common name applied to them, remains to be determined.

6.5 Development of the Posterior Hypothalamus: Dorsal Region

The components of the dorsal region of the posterior hypothalamus derive from the corresponding neuroepithelium (phd), which in the midline third ventricle of young embryos is separated from the ventral region of the posterior hypothalamic rudiment (phv) by the mammillary neuroepithelium (ma) (Figs. 3A, B, 7A, B). More laterally, where the mammillary recess is not present, the two components of the posterior hypothalamic neuroepithelium are contiguous but still distinguishable from each other (Figs. 4A, B, 8A, B). Dorsally the neuroepithelium of the dorsal region of the posterior hypothalamus is contiguous with the upper tegmental neuroepithelium (teu).

6.5.1 The Supramammillary Nucleus

Background. The supramammillary nucleus forms a cap over the mammillary body, but it does not quite extend to either the rostral or caudal boundary of the mammillary body. It is composed of small- and medium-sized cells. The cells situated laterally tend to be elongated in the transverse plane, paralleling fibers that traverse this region; the medial cells tend to be roundish and are less densely packed (Fig. 93). On the basis of these criteria we distinguish between medial (SMm) and lateral (SMl) parts of the supramammillary nucleus. Some of the cells of the supramammillary nucleus have been described to show cholecystokinin-like immunoreactivity (Greenwood et al. 1981). The efferents of the supramammillary nucleus distribute to the hippocampus (Segal and Landis 1974; Pasquier and Reinoso-Suarez 1976), particularly the dentate gyrus (Segal 1979; Haglund et al. 1984), and other limbic structures, such as the nucleus accumbens and septum (Carter and Fibiger 1977). It receives afferents from structures that also distribute to the mammillary body, the preoptic area, the lateral hypothalamus, and the septum (Swanson and Cowan 1979; Saper et al. 1979), and from the raphe nuclei (Taber Pierce et al. 1976; Bobilier et al. 1979). Our earlier quantitative radiographic study has shown that peak production of supramammillary neurons occurs on day E16, and that neurogenesis comes to an end on day E17 (Altman and Bayer 1978a: Fig. 9).

Results. An examination of thymidine radiograms of rats that received [^{3}H]thymidine between days E14 and E19 and were killed on day E22 (shown in the context of the development of the mammillary body in Figs. 101–103) indicates the following about the time of origin of supramammillary neurons (SM). In rats injected on day E14, the neurons are very lightly labeled (Fig. 101A). In rats injected on day E15 (Fig. 101B) there is intermediate labeling, with a few heavily labeled cells. This suggests that few supramammillary neurons are produced before day E15. In rats injected on day E16 many of the pars medialis cells (SMm in Fig. 102A) are no longer labeled and others show heavy labeling; in the pars lateralis (SMl) most cells are either heavily or lightly labeled. In rats injected on day E17 (Fig. 102B) only a rare cell is labeled medially but a fair proportion is still labeled laterally. The lateral cells are generally no longer labeled in rats injected on day E18 (Fig. 103A) or day E19 (Fig. 103B). These observations indicate that there is a medial-to-lateral gradient in the

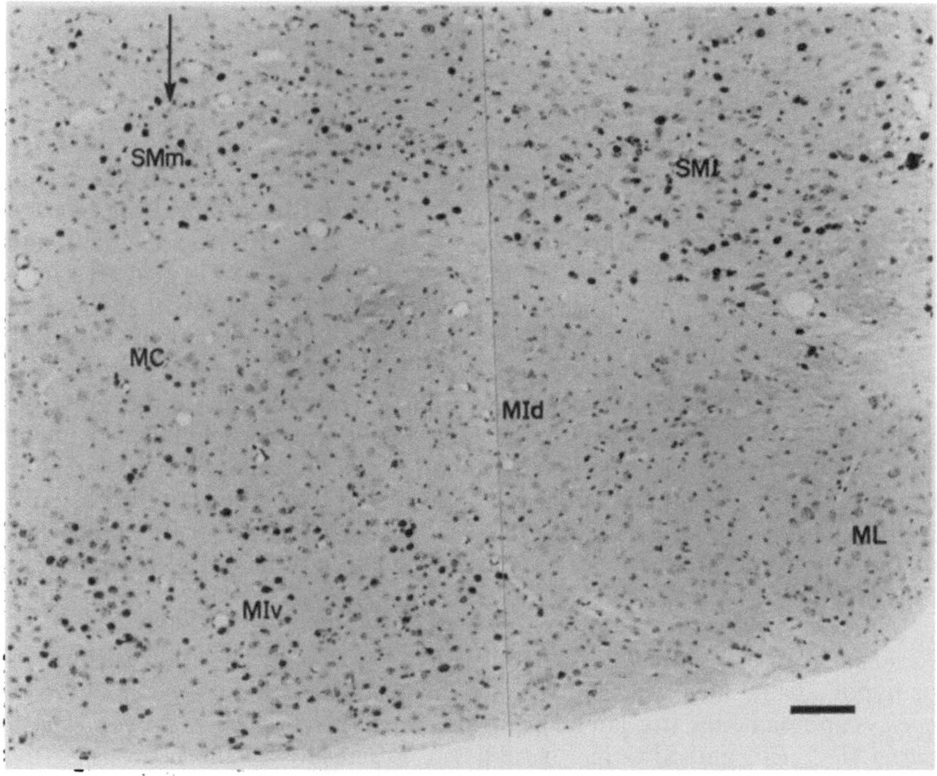

Fig. 93. Coronal thymidine radiogram from an adult rat tagged on days E16+17, showing the two components of the supramammillary nucleus in relation to the nuclei of the mammillary body. Paraffin. *Scale*, 100 μm

settling of supramammillary neurons. The neurons of the pars medialis are generated predominantly on days E15 and E16, and the neurons of the pars lateralis on days E16 and E17 (compare with Fig. 93).

Comments. The development of the supramammillary nucleus has received little attention in the past. The present results suggest that in spite of its contiguity with the mammillary nuclei, it is not a part of the mammillary body because its neurogenetic gradient (medial-to-lateral) is the reverse found in all components of the mammillary body (Sect. 6.6). It is likely, therefore, that it does not originate in the neuroepithelium of the mammillary recess, which is the source of the neurons of all of the mammillary nuclei. The recent demonstration of a projection from the supramammillary nucleus to the dentate gyrus of the hippocampus (Segal 1979; Haglund et al. 1984) suggests, however, that functionally the supramammillary nucleus may be part of a loop in which the mammillary nuclei receive a major input from the subicular region of the hippocampal formation (Meibach and Siegel 1977; Swanson and Cowan 1977), and the supramammillary nuclei provide a return path directly to the hippocampus.

6.5.2 The Posterior Area Nuclei

Background. Gurdjian (1927) and Krieg (1932) have described the posterior hypothalamic nucleus as a somewhat diffuse midline structure situated between the dorsomedial nucleus anteriorly and the periventricular gray of the midbrain posteriorly. Their Golgi studies suggested that axons from cells in this region reach the periventricular area of the hypothalamus. Other authors have referred to this region as the posterior hypothalamic area (e.g., Bleier et al. 1979), and Christ (1969) uses the term dorsocaudal hypothalamic area. The scattered indirect studies (reviewed by Palkovits and Záborszky 1979) suggest that this area receives afferents from the septum and the hippocampus by way of the fornix and the medial forebrain bundle, and from the midbrain central gray by way of the dorsal longitudinal fasciculus. It sends efferents to the central gray, the locus coeruleus, and the spinal cord. This region has been implicated in the regulation of ACTH secretion (Grizzle et al. 1974). Our quantitative thymidine radiographic investigation in adult rats showed that the neurons of the posterior hypothalamic area are generated between days E13 and E15, with about 70% of the cells being produced on days E14 and E15 (Altman and Bayer 1978a: Fig. 8).

Results. Because we were able to identify three separate nuclei within the posterior hypothalamic area, we shall refer to the central portion as the *dorsocaudal nucleus* and name the two lateral structures as the *periformical* and *supraformical nuclei* (DC, PF, and SFO in Figs. 89, 90). Practically all the neurons of the dorsocaudal nucleus are lightly or heavily labeled in perinatal rats injected on day E14 (not shown). In perinatal rats injected on day E15 there is an admixture of unlabeled and heavily labeled cells (Fig. 89A) but following injections on days E16 or E17 (Figs. 89B, 90) only occasional cells are labeled. This indicates that the bulk of the dorsocaudal nucleus neurons are produced on days E14 and E15. There is no indication of a neurogenetic gradient within the dorsocaudal nucleus.

Lateral to the dorsocaudal nucleus and overlying the descending fornix is the supraformical nucleus (SFO). Many of its neurons are heavily labeled in perinatal rats injected on day E15 (Fig. 89A). In its ventral half, heavily labeled cells persist in rats injected on days E16 (Fig. 89B) and E17 (Fig. 90). This indicates that the neurons of the supraformical nucleus, as a whole, are produced somewhat later than the neurons of the dorsocaudal nucleus and that, unlike in the latter, there is a neurogenetic gradient (dorsal-to-ventral; arrows in Fig. 89B) in this region.

In the periformical nucleus (PF) the proportion of labeled cells is lower than in the two previous structures in perinatal rats injected on day E14 (Fig. 89A), and only a few labeled cells are present in rats injected on day E15 (Fig. 89B). It is presumed that these three nuclei stem from a neuroepithelial zone dorsal to the site of origin of the premammillary nuclei (Figs. 91, 92).

Comments. We were able to distinguish in radiograms from perinatal rats, by virtue of their labeling patterns, three components of the posterior hypothalamic area. The dorsocaudal nucleus bridges the midline. There is no neurogenetic gradient here, presumably because the cells of the nucleus are just "left behind" as the ventricular lining is dissolved. But a gradient could be detected in the suprafornical nucleus, suggesting a dorsolateral migratory path for its cells. There is a sequential order in the production of neurons in the three posterior hypothalamic nuclei: the perifornical nucleus first, the rostrocaudal nucleus next, and the suprafornical nucleus (at least a part of it) last.

6.6 Development of the Mammillary Body

Background. The mammillary body is subdivided by all investigators into two major components, the smaller lateral nucleus and the larger medial nucleus. However, there is divergence of opinion about the parcelling of the medial mammillary nucleus, which is further complicated by the circumstance that considerable species differences exist in the location, shape, and size of its parts or subnuclei (Rose 1939). In the rat, Gurdjian (1927) distinguished six parts in the medial nucleus: the pars medianus, pars medialis, pars lateralis, pars posterior, pars commissuralis dorsalis, and pars commissuralis ventralis. Krieg (1932) reduced these to three: pars medialis, pars lateralis, and pars posterior, admitting the possibility that the pars medialis might be divisible into a dorsal and a ventral component. In a recent study, Seki and Zyo (1984) distinguished four subnuclei in the medial mammillary nucleus of the rat: the pars medialis centralis, pars medialis dorsalis, pars medialis lateralis, and pars basalis. With the exception of the pars basalis, the subdivisions of the medial mammillary nucleus are conceptualized by these investigators as vertical columns, which agrees with the impression one gets at all levels except posteriorly, where the medial mammillary nucleus has a horizontally directed oval shape.

The major afferents of the mammillary body are from the subiculum (Swanson and Cowan 1975, 1977; Meibach and Siegel 1975, 1977) and the septum (Meibach and Siegel 1977) by way of the fornix and the medial forebrain bundle, and from the tegmentum by way of the mammillary peduncle (Nauta and Kuypers 1958; Cowan et al. 1964; Fry and Cowan 1972). The ascending outflow is to the anterior thalamic nuclei by way of the mammillothalamic tract; the distribution of these fibers is topographically organized (Seki and Zyo 1984). The targets of the descending fibers in the mammillotegmental tract are the dorsal and ventral tegmental nuclei and regions of the pons (Morin 1950; Cruce 1977).

Our earlier quantitative radiographic studies established that the neurons of the lateral mammillary nucleus are generated between days E12 and E15, with a peak on day E13 (Altman and Bayer 1978a: Fig. 9). The pattern of labeling in the late-produced components of the medial mammillary nucleus suggested a horizontal rather than a vertical parcellation (Altman and Bayer 1978a: Fig. 22). We have tentatively distinguished a medial portion of the nucleus with peak production of neurons on day E15, and a principal portion (with a central and ventral subdivision) with peak production on day E16. In light of the controversies referred to, we have reexamined the organization of the mammillary nucleus in serial sections applying cytological criteria. We will relate the radiographic patterns during embryonic development as well as in adults to this new subdivision.

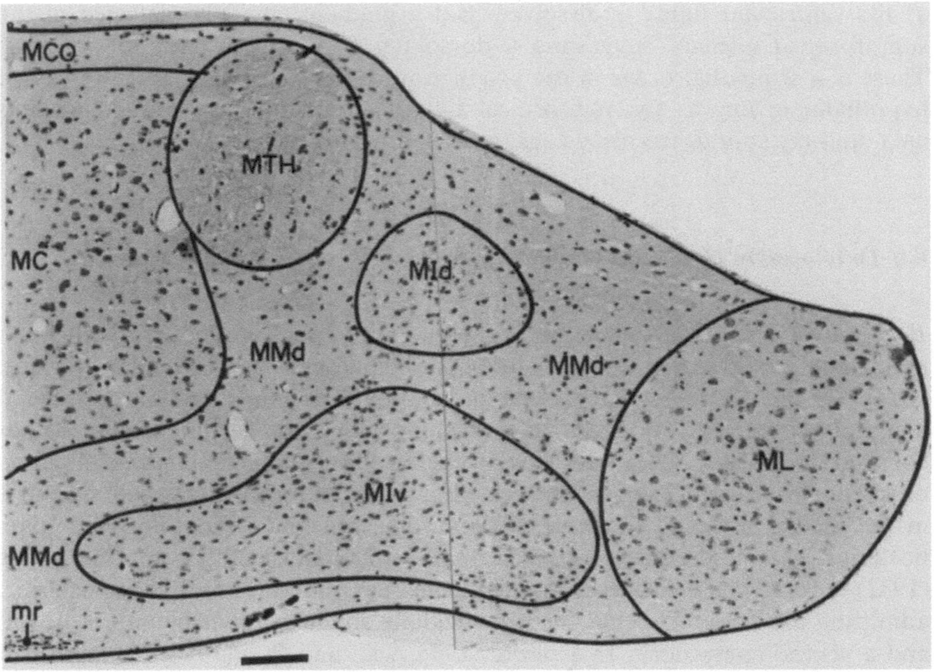

Fig. 94. Coronal section through the mammillary body of an adult rat to illustrate the cytological criteria used for the subdivision of the mammillary body. Paraffin. *Scale*, 100 μm

Results. The criteria used for the subdivision of the mammillary body are illustrated in Fig. 94. They included a clustering of cells against a cell-sparse, or diffusely organized, background, and the size and orientation of the cells within the clusters. The *lateral nucleus* (ML) is clearly demarcated from the medial nucleus by its far-lateral position and its large cells. In the *medial nucleus*, embedded in a cell-sparse matrix, referred to as the diffuse mammillary region (MMd), three cell clusters are distinguished at this level (compare with Fig. 95B). The *central* subnucleus (MC) is situated dorsomedially and is composed of cells that are somewhat smaller than those of the lateral nucleus. The *intermediate* subnucleus is divided into two parts, a *pars ventralis* (MIv), composed of small, horizontally oriented spindle-shaped cells, and a *pars dorsalis* (MId) composed of similar small cells that appear roundish in coronal sections. The shape and size of these components of the mammillary body vary from rostral to caudal (Fig. 95; right half of panel). The lateral mammillary nucleus is most extensive rostrally (Fig. 95A), tapers off caudally (Fig. 95B–E), and does not reach the most caudal portion of the mammillary body (see Fig. 97D). The central subnucleus varies in size and shape from rostral to caudal, perhaps influenced by the size and shape of the mammillothalamic tract (MTH). The pars ventralis of the intermediate subnucleus is larger caudally than rostrally and, in some rats, its two halves fuse medially as the mammillary body shrinks caudally. The size and shape of the pars dorsalis of the intermediate subnucleus varies in shape at different levels, and at several sites it is difficult to delineate from the *diffuse mammillary region*. The latter is most extensive rostrally.

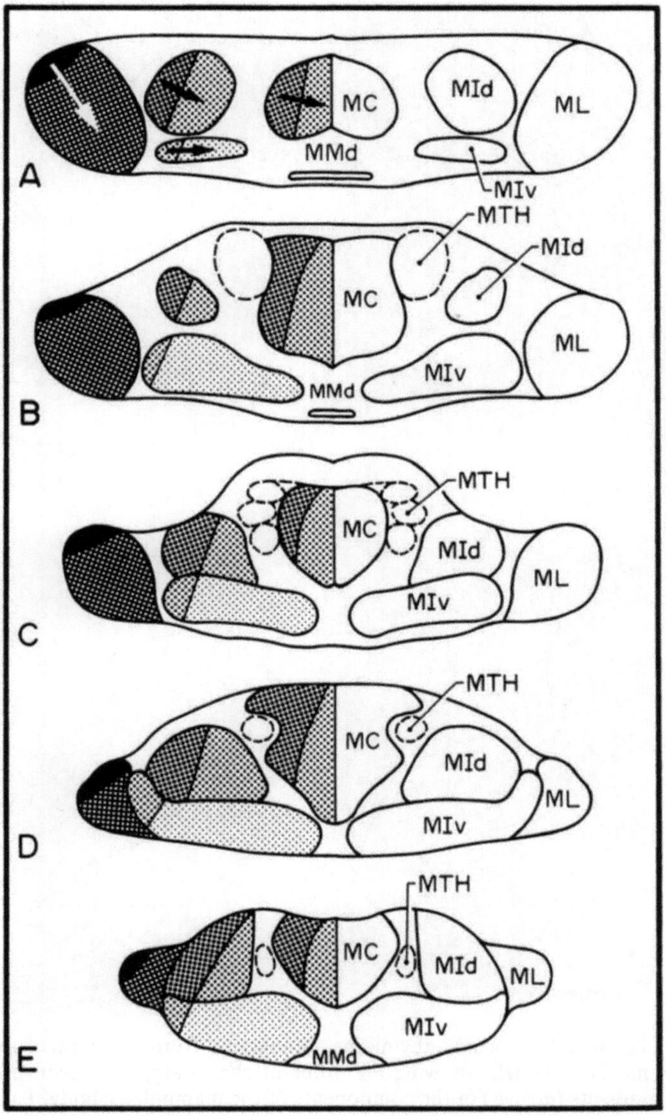

Fig. 95 A–E. Tracings of coronal sections of the mammillary body of an adult rat from rostral (**A**) to caudal (**E**). Neurogenetic gradients within the subdivisions of the mammillary body are indicated *on the left*. *Shadings* reflect the chronology of neurogenesis from earliest (*opaque*) to latest (*fine dots*). *Arrows* in **A** indicate neurogenetic gradients

In adult rats injected on day E15+16 some of the neurons of the lateral mammillary nucleus are still labeled (not shown), but few if any are labeled in rats injected on days E16+17 (Fig. 96A). In the central subnucleus the laterally situated cells are unlabeled in this group and the medially situated cells tend to be labeled, indicating a lateral-to-medial neurogenetic gradient (arrow in Fig. 96A). The same pattern is seen in the pars dorsalis of the intermediate subnucleus. This indicates two separate lateral-to-medial gradients in the dorsal portion of the medial nucleus: one in the central subnucleus and another in

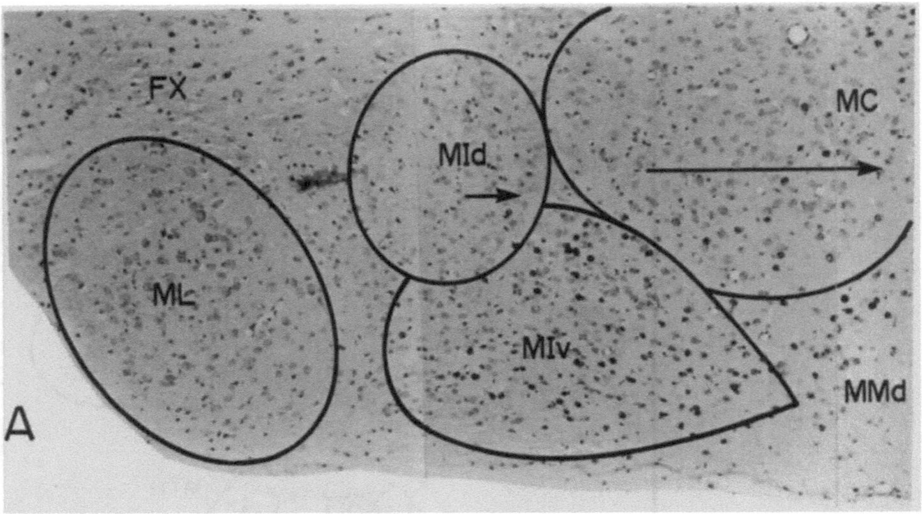

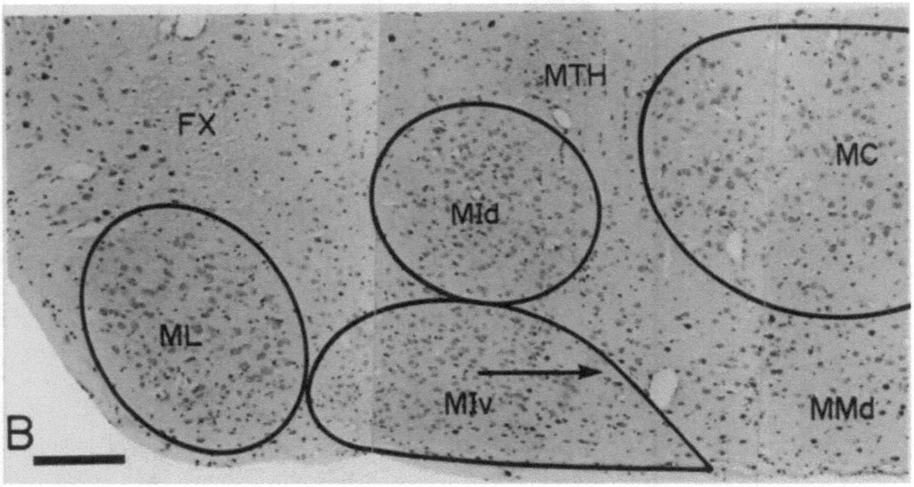

Fig. 96A, B. Coronal thymidine radiograms from adult rats labeled on days E16+17 (**A**) and E17+18 (**B**), showing the different chronology of neuron production, and neurogenetic gradients (*arrows*) in the components of the mammillary body. Paraffin. *Scale*, 150 μm

the pars dorsalis of the intermediate subnucleus. The pars ventralis of rats injected on days E16+17 is composed of labeled cells throughout (Fig. 96A). In rats injected on days E17+18 (Fig. 96B), labeled cells are restricted to the medial portion of the pars ventralis of the intermediate subnucleus and to the medial part of the diffuse region. This indicates that ventrally there is a single lateral-to-medial gradient in the medial mammillary nucleus (arrow in Fig. 96B). This pattern of neurogenetic gradient appears throughout the mammillary body from rostral to caudal (Fig. 97) without any evidence of a longitudinal neurogenetic gradient.

The neurons of the mammillary body originate in the neuroepithelium of the mammillary recess (ma), which is a clearly recognizable subdivision of the lining of the third ventricle by day E13 (Fig. 3). As seen in a rat injected on

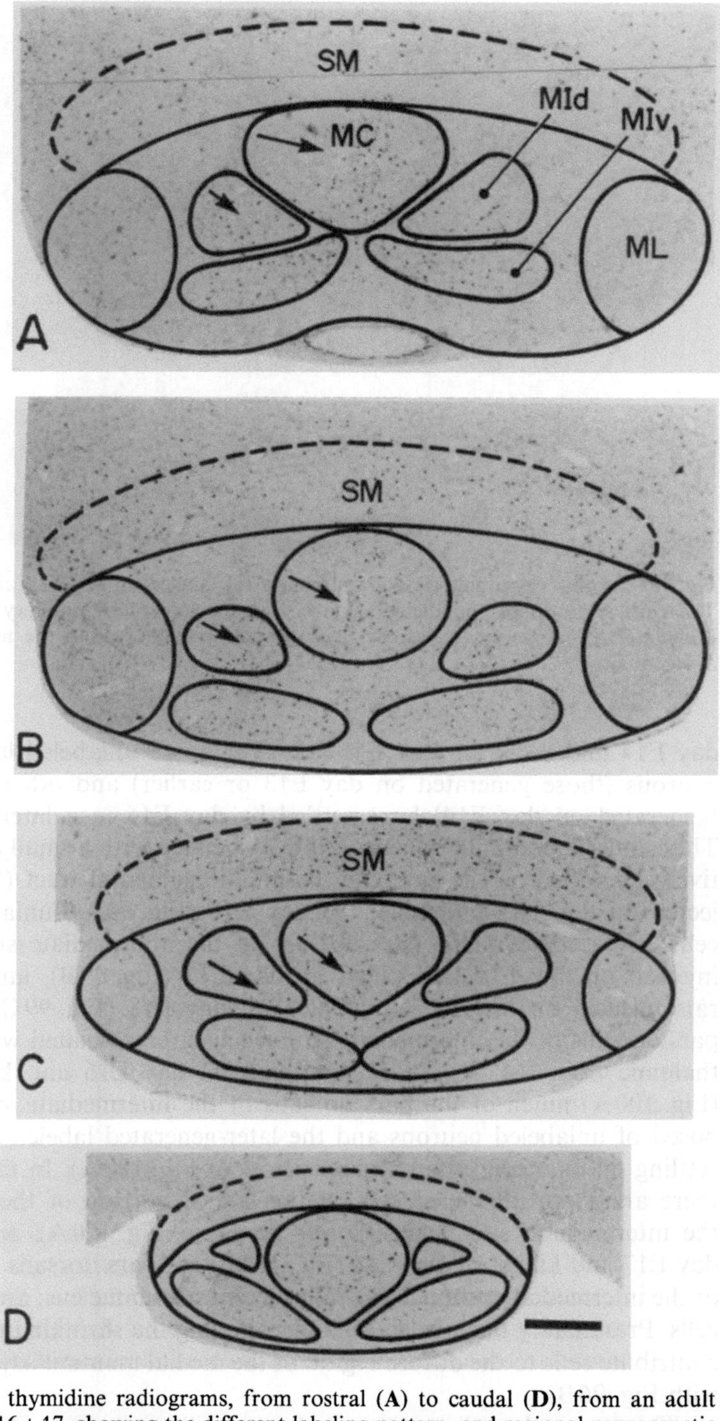

Fig. 97 A–D. Coronal thymidine radiograms, from rostral (**A**) to caudal (**D**), from an adult rat labeled on days E16 + 17, showing the different labeling pattern, and regional neurogenetic gradients (*arrows*) in the subdivisions of the mammillary body. Paraffin. *Scale*, 300 µm

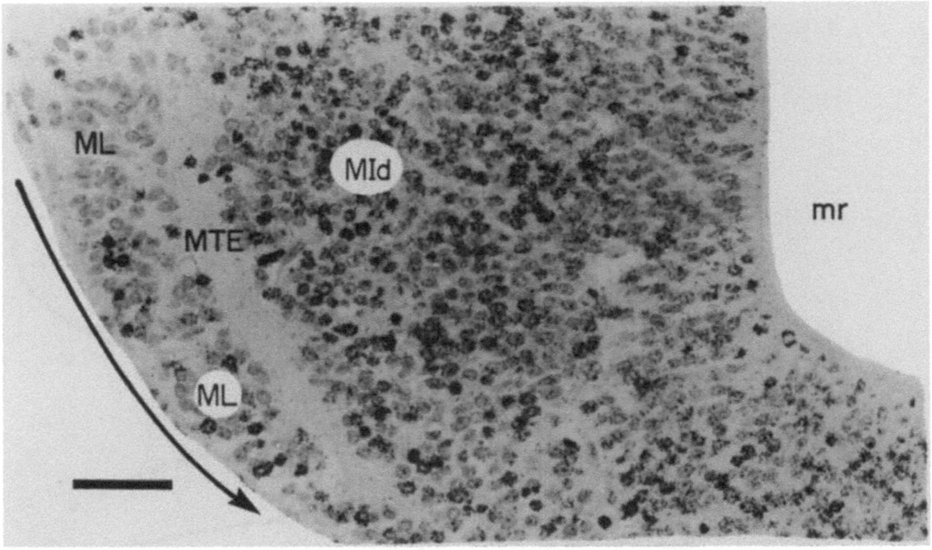

Fig. 98. Coronal thymidine radiogram from a rat labeled on day E14 and killed on day E15. The earliest-produced (unlabeled) neurons on the lateral mammillary nucleus have settled dorsolaterally. *Arrow* indicates the neurogenetic gradient within the nucleus. Methacrylate. *Scale*, 50 μm

day E14 and killed on day E15 (Fig. 98), some unlabeled lateral mammillary neurons (those generated on day E13 or earlier) and others that are labeled (generated on day E14) have settled by day E15 in a lateral position (ML). These lateral mammillary neurons are associated with a small fiber bundle tentatively identified as the emerging mammillotegmental tract (MTE). In rats injected on day E15 and killed on day E16 (Fig. 99A), unlabeled and labeled cells have settled in the pars dorsalis of the intermediate subnucleus. In rats injected on day E15 and killed on day E17 (Fig. 99B), and more clearly in rats injected on day E15 and killed on day E18 (Fig. 99C), the cells of the pars dorsalis of the intermediate subnucleus are associated with the mammillothalamic tract (MTH). In rats injected on day E16 and killed on day E17 (Fig. 100A) much of the pars dorsalis of the intermediate subnucleus is composed of unlabeled neurons and the later-generated labeled cells appear to be settling on its medial flank (upper arrow in Fig. 100A). In this injection group there are some unlabeled cells in the lateral portion of the pars ventralis of the intermediate subnucleus (lower arrow in Fig. 100A). In rats injected on day E17 and killed on day E18 (Fig. 100B) the pars dorsalis and pars ventralis of the intermediate subnucleus, and the central subnucleus, are devoid of labeled cells. Presumably the labeled cells surrounding the shrinking mammillary recess contribute cells to the diffuse region of the medial mammillary nucleus (compare with Fig. 96B).

The terminal settling pattern of mammillary neurons is illustrated in rats injected between days E14 and E19 and killed the day before birth on day E22 (Figs. 101–103). Many of the neurons of the lateral nucleus are heavily labeled in the animals injected on day E14 (Fig. 101A) with a few dorsolaterally situated cells being unlabeled (compare with Fig. 98A). In animals injected on

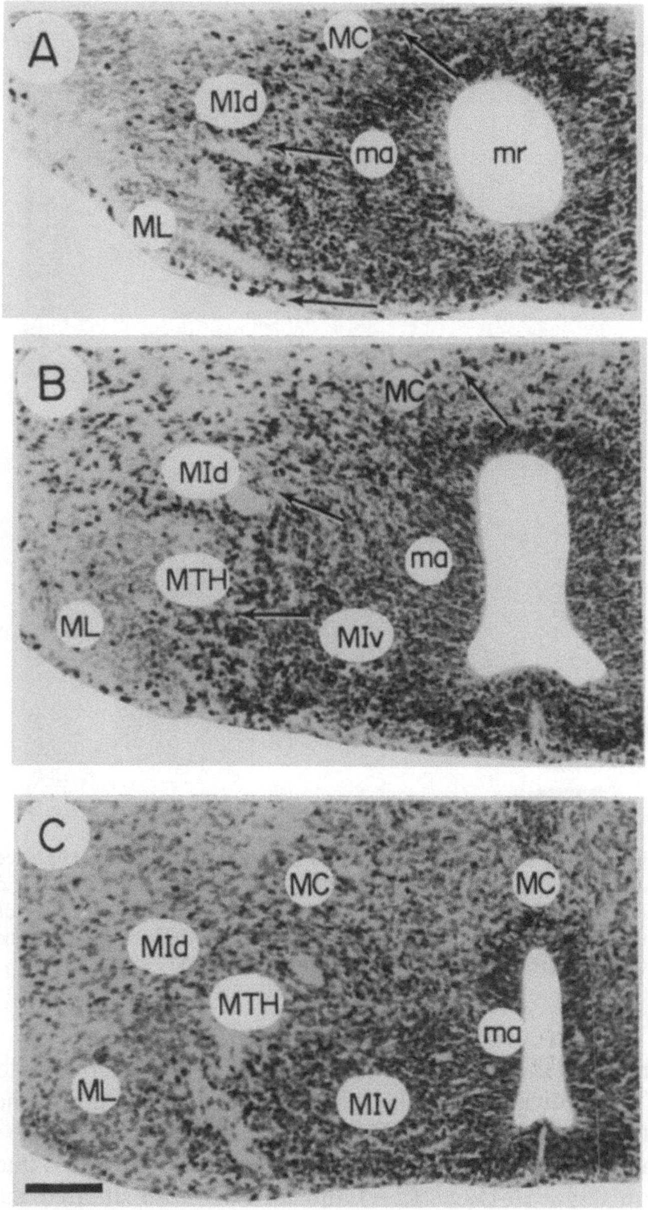

Fig. 99 A–C. Coronal thymidine radiograms from rats labeled on day E15 and killed on day E16 (**A**), E17 (**B**), and E18 (**C**) showing the apparently radial migratory route of neurons (*arrows*) to the proximal part of the lateral mammillary nucleus and distal parts of components of the intermediate and central subnuclei. Paraffin. *Scale*, 100 μm

day E15 (Fig. 101 B), the few labeled cells in the lateral nucleus are concentrated ventromedially. This indicates a dorsolateral-to-ventromedial gradient within the lateral mammillary nucleus. In the dorsal portion of the intermediate subnucleus heavily labeled cells predominate in rats injected on day E15 (Fig. 101 B); a few cells are unlabeled laterally. In rats injected on day E16 (Fig. 102 A)

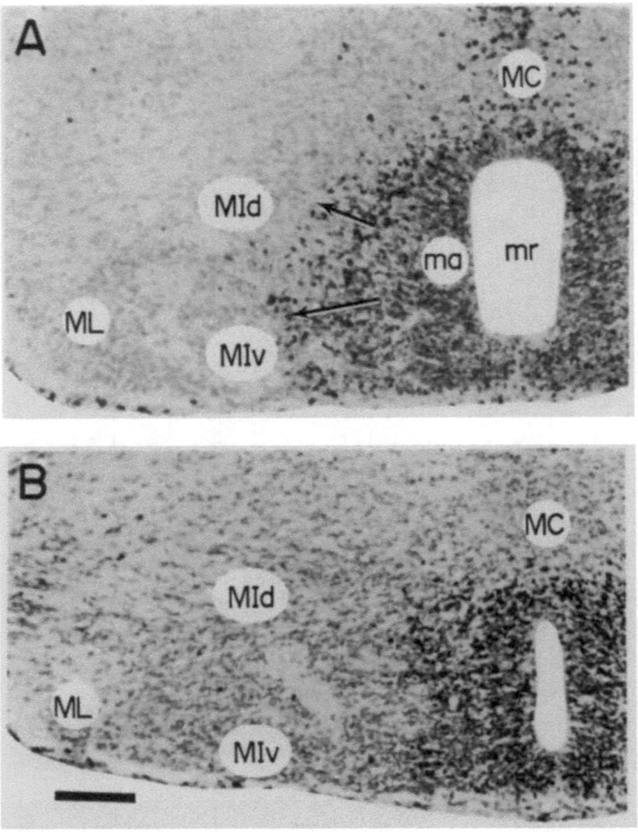

Fig. 100 A, B. Coronal thymidine radiograms from a rat labeled on day E16 and killed on day E17 (**A**), and another labeled on day E17 and killed on day E18 (**B**). In the former, labeled cells are approximating the lateral flanks of the pars dorsalis and pars ventralis of the intermediate subnucleus (*arrows in* **A**); in the latter, labeled cells are limited to the vicinity of the neuroepithelium. Paraffin. *Scale*, 100 μm

none of the cells of the pars dorsalis of the intermediate subnucleus are labeled. This indicates that this region comprises the next generation of neurons after the lateral mammillary neurons, and the gradient here is lateral-to-medial. The unlabeled cells of the central subnucleus are located laterally in the rats injected on day E16 (Fig. 102A) and all but those situated medially are unlabeled by day E17 (Fig. 102B). The central subnucleus is thus composed of neurons that are produced third in sequence, and these, too, settle in a lateral-to-medial pattern. Heavily labeled cells appear in a dorsolateral position in the ventral part of the intermediate subnucleus on day E16 (Fig. 102A), while in rats injected on day E17 the heavily labeled cells are restricted to a ventromedial band (Fig. 102B). Few if any cells of this region are labeled in rats injected on day E18 (Fig. 103A). The neurons of the pars ventralis of the intermediate subnucleus are the fourth in order of production and they settle in a dorsolateral-to-ventromedial sequence. The only cells labeled in rats injected on day E18 are those belonging to the diffuse region of the medial mammillary nucleus (MMd in Fig. 103A). Some of these cells are still labeled on day E19 (Fig. 103B).

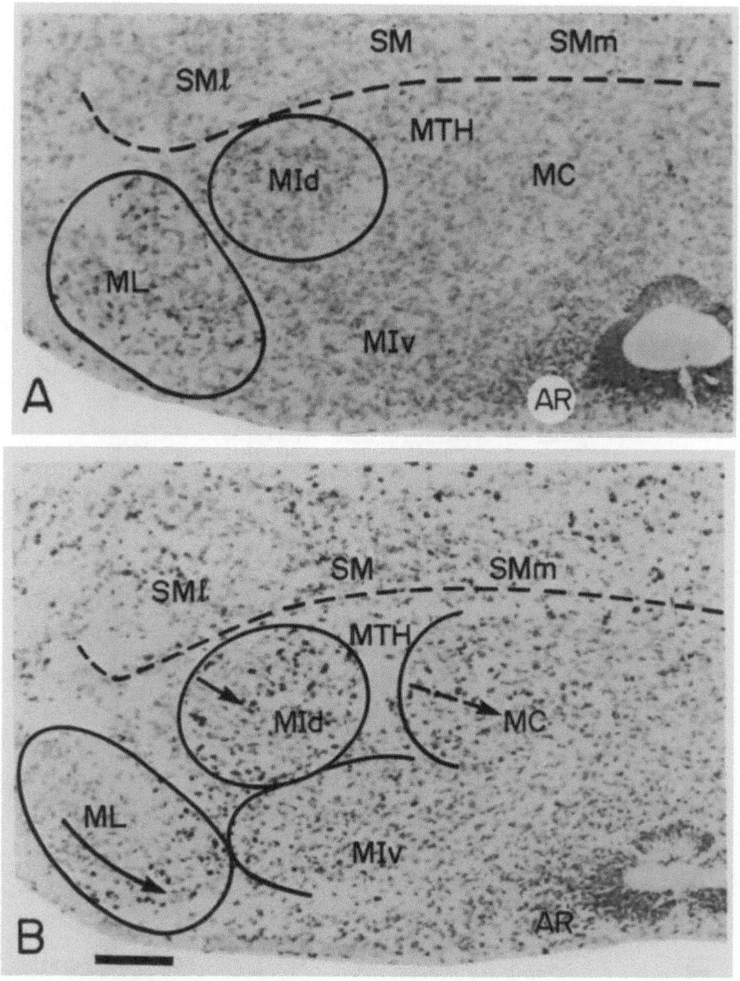

Figs. 101–103. Coronal thymidine radiograms from perinatal rats tagged on day E14 (Fig. 101 A), E15 (Fig. 101 B), E16 (Fig. 102 A), E17 (Fig. 102 B), E18 (Fig. 103 A), and E19 (Fig. 103 B), showing the chronology of neurogenesis and neurogenetic gradients in the midportion of the mammillary body. Paraffin. *Scale*, 100 μm

This region represents the last-generated components of the mammillary nucleus.

Comments. We have subdivided the medial mammillary nucleus on the basis of cytological criteria into a central and an intermediate subnucleus. The central nucleus is composed of relatively large cells (somewhat smaller than the cells of the lateral mammillary nucleus) and the intermediate subnucleus of small cells. The latter was further divided, partly on the basis of the orientation of the cells, into a pars dorsalis and a pars ventralis. These three components of the medial mammillary nucleus vary in shape and size from rostral to caudal, as does the diffuse region of the nucleus in which these longitudinal cell columns are embedded (Fig. 95). There is a sequential order in the production of mammillary neurons. The lateral mammillary neurons are produced first and its

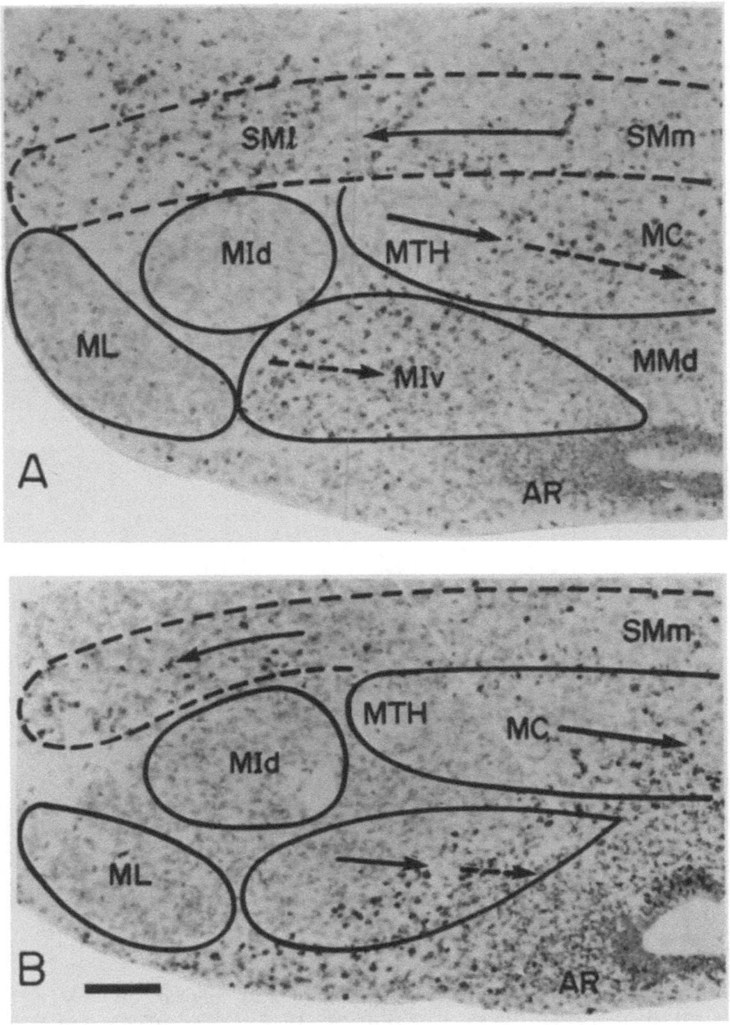

Fig. 102. Legend see Fig. 101

latest-produced cells (day E15) are situated ventromedially. The neurons of the central subnucleus and of the pars dorsalis of the intermediate subnucleus are next, and in each the latest cells (day E16) are situated medially. Most of the neurons of the pars ventralis of the intermediate subnucleus are produced on day E16 and the last produced cells (day E17) are again located medially. The latest elements of the mammillary body (days E18 and E19) are located in the ventromedial portion of the diffuse region of the medial nucleus.

The settling of mammillary neurons follows a modified lateral-to-medial (or outside-in) pattern that does not hold true for the mammillary body as a whole, only for the subdivisions (compare with left panel in Fig. 95). Thus the medially situated neurons of the pars dorsalis of the intermediate subnucleus are produced later than the more medially situated neurons of the lateral portion of the central subnucleus (Fig. 97A, B). This pattern suggests that each subnucle-

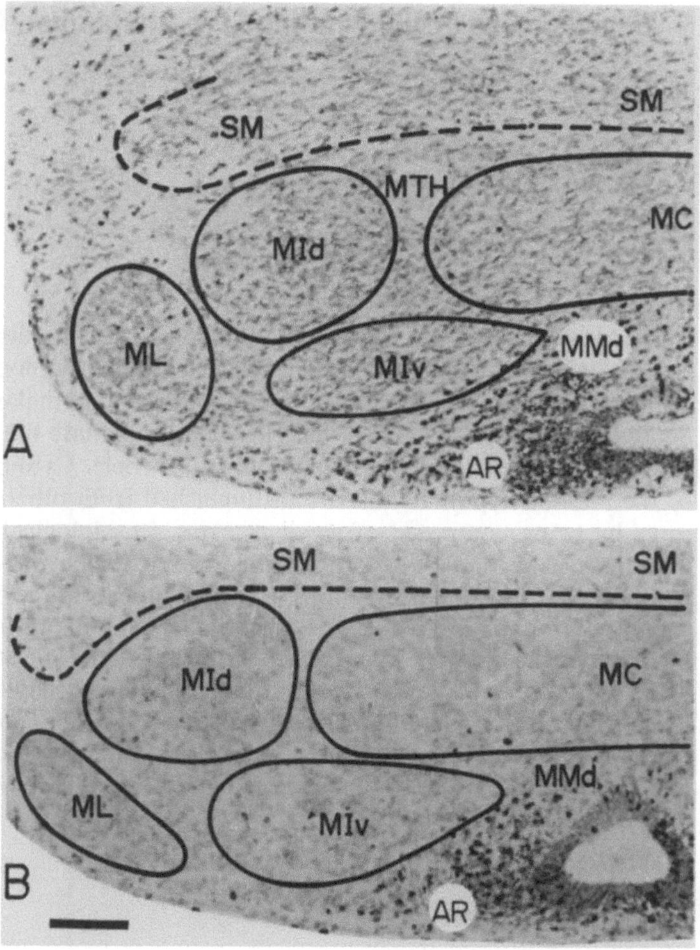

Fig. 103. Legend see Fig. 101

us is derived from a different germinal zone of the mammillary neuroepithelium each with its own chronology and outside-in pattern of settling. The arrows illustrating the regional patterns of the settling of neurons (Fig. 99 A, B) suggest that the lateral mammillary nucleus derives from the basal portion of the mammillary recess neuroepithelium, the intermediate subnucleus from its intermediate part, and the central subnucleus from its dorsal part.

7 Development of the Midline Hypothalamus: The Third Wave

The components of the midline hypothalamus can be distinguished from the rest of the hypothalamus by topographical, cytological, physiological, and developmental criteria. Topographically, the midline hypothalamus consists of condensed and diffuse structures that closely approximate the ependymal and specialized linings of the hypothalamic third ventricle. Cytologically, components of the midline hypothalamus are distinguished from other hypothalamic structures by being composed of small neurons. Physiologically, the parvocellular neurons of the midline hypothalamus appear to be directly involved in the regulation of hormonal functions, and may constitute a major portion of the "hypophysiotropic" area of the hypothalamus. Finally, developmentally, the neurons of the midline hypothalamus represent the latest-generated, or third wave, of neurons produced by the hypothalamic neuroepithelium. The midline hypothalamus consists of two specialized structures situated in a ventral position, the suprachiasmatic nucleus anteriorly and the arcuate nucleus posteriorly, and of the diffuse periventricular field that extends all the way from the preoptic area to the mammillary body.

7.1 The Suprachiasmatic Nucleus

Background. The suprachiasmatic nucleus is identifiable by its location and its singularly small and densely packed cells. It has been proposed (van den Pol 1980) that it is composed of two parts, a ventrolateral and a dorsomedial group of cells. The cells are somewhat less densely packed ventrolaterally and this is the region where afferent terminals from several structures are concentrated. These afferents include retinal fibers (Moore and Lenn 1972; Hendrickson et al. 1972; Mai and Junger 1977), fibers from the ventral lateral geniculate nucleus (Swanson et al. 1974; Ribak and Peters 1975), and serotonergic fibers from the raphe nuclei (Fuxe 1965; Aghajanian et al. 1969). The somewhat less densely packed cells of the dorsomedial region contain vasopressin and neurophysin (Vandesande et al. 1975; Sofroniew and Weindl 1978). It has been reported that the fine-caliber vasopressin- and neurophysin-containing efferents reach a variety of extrahypothalamic and hypothalamic structures but not the median eminence (Sofroniew and Weindl 1978; Buijs 1978). However, the majority of Golgi-impregnated cells were found to have locally terminating axons (van den Pol 1980). The subdivision of the suprachiasmatic nucleus into a ventrolateral and dorsomedial component agrees with the neurogenetic difference observed in adult rats with long-survival thymidine radiography. The neurons of the suprachiasmatic nucleus are produced between days E14 and E17 with a peak on day E15 (Altman and Bayer 1978a: Fig. 5). The earlier-generated cells are concentrated ventrolaterally and the later-generated cells dorsomedially (Altman and Bayer 1978a: Fig. 10).

Results. In rats labeled with [^{3}H]thymidine on day E15 and killed on day E22 (Fig. 104) the earlier-produced cells situated ventrolaterally (the *ventrolateral suprachiasmatic subnucleus*; SUv) are heavily labeled while the later-produced cells situated dorsomedially (the *dorsomedial suprachiasmatic subnucleus*; SUd) are lightly labeled. In rats tagged on day E16 and killed on day E22, most of the cells of the ventrolateral subnucleus are no longer labeled but those of the dorsomedial subnucleus are labeled (not shown). In rats labeled on day E17 and killed on day E22 (Fig. 105) some of the cells of the dorsomedial subnucleus are still labeled. The concentration of these heavily labeled cells is higher rostrally (Fig. 105A) than caudally (Fig. 105B). In addition a narrow band of cells in the caudal portion of the suprachiasmatic nucleus adjacent to the optic chiasma ventromedially and ventrally contain several heavily labeled cells; this region is designated as the *basal suprachiasmatic subnucleus* (SUb in Fig. 105B, C). These observations indicate that the neurons of the ventrolateral suprachiasmatic nucleus are produced first (mainly on day E15), those of the dorsomedial suprachiasmatic nucleus next (on days E16 and E17), and those of the basal suprachiasmatic nucleus last (day E17).

In our previous study with long-survival thymidine radiography we proposed that in addition to a ventrolateral-to-dorsomedial neurogenetic gradient in the adult suprachiasmatic nucleus (Fig. 106B) there is also a rostral-to-caudal gradient (Fig. 106A). Our present investigation shows that within the two major subnuclei of the suprachiasmatic nucleus there is no such gradient. In fact, in the dorsomedial subnucleus a caudal-to-rostral gradient is indicated (Fig. 105). The concentration of heavily labeled cells caudally with late injections is, accordingly, reinterpreted to represent the basal suprachiasmatic subnucleus (Figs. 105B, C, 106A).

The neurons of the suprachiasmatic nucleus are derived from a special neuroepithelial zone. The suprachiasmatic neuroepithelial zone (su), illustrated in short-survival radiograms in day E15 (Fig. 107) and day E16 rats (Fig. 108) in two different frontal planes, forms a shallow ventromedian sulcus caudal to the optic recess (Fig. 107A). This neuroepithelial zone is divisible into two parts, tentatively designated as the dorsomedial (sud) and ventrolateral (suv) suprachiasmatic primordia (Figs. 107B, 108B). By day E18 (Fig. 109) this sulcus has decreased in depth and the region is occupied by the expanding optic chiasma.

The clustering of differentiating cells of the suprachiasmatic nucleus is not evident on day E15 (Fig. 107B) but is obvious by day E16 (SUv in Fig. 108). The mostly unlabeled differentiating cells in rats labeled on day E16 and killed 2 h later are considered to represent the ventrolateral subnucleus. Its cells may derive from the less-active ventrolateral neuroepithelium (suv in Fig. 108B). The labeled cells within the differentiating subnucleus represent locally proliferating cells, presumably glial cells rather than neurons. This is supported by the observation that cells in the same location are still labeled in rats injected on day E18 and killed 2 h later (Fig. 109B), that is, after the cessation of the production of suprachiasmatic neurons (Altman and Bayer 1978a: Fig. 5). The

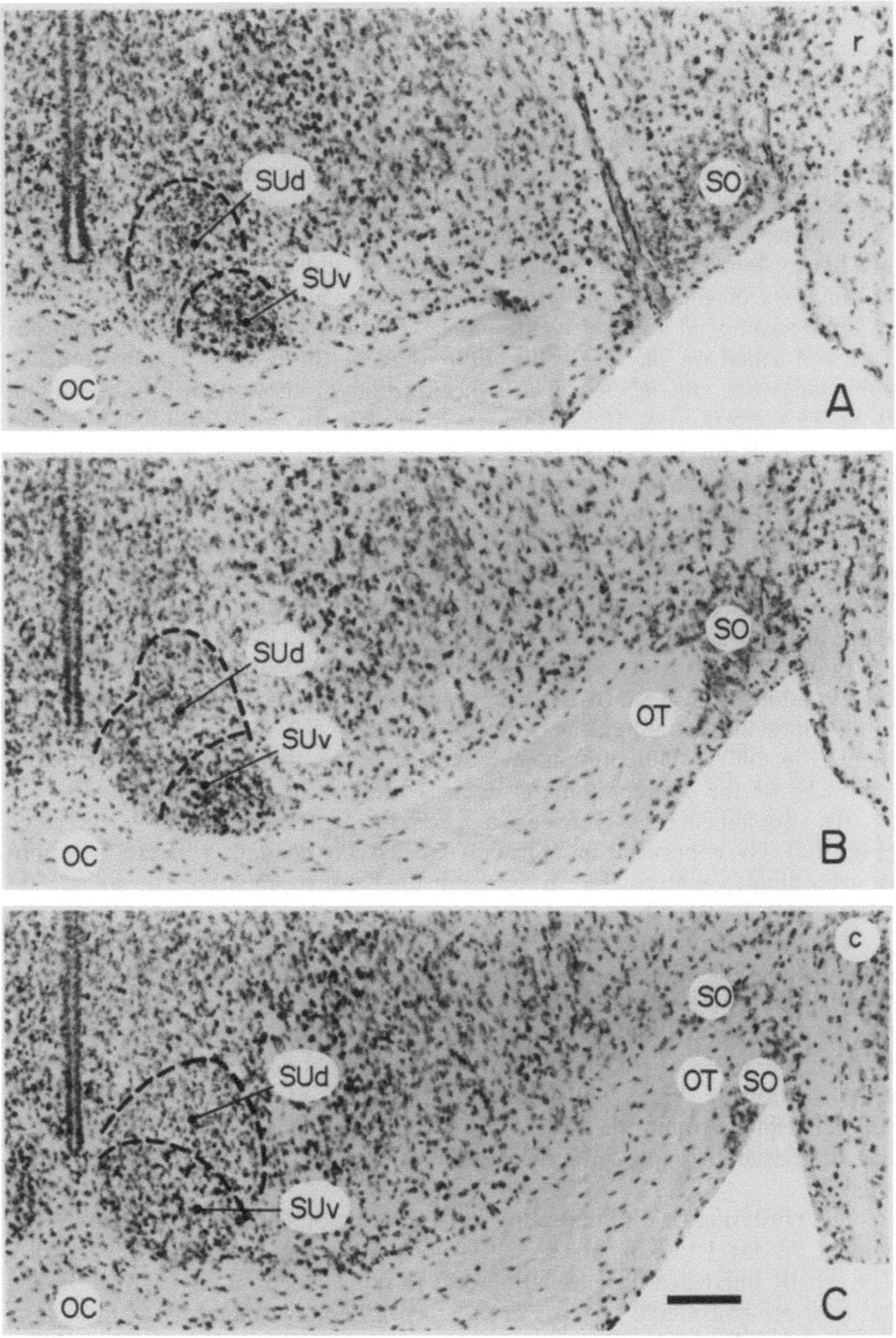

Fig. 104 A–C. Coronal thymidine radiograms, from rostral (**A**) to caudal (**C**), from a perinatal rat labeled on day E15. The ventrolateral suprachiasmatic subnucleus is composed of heavily labeled cells, the dorsomedial suprachiasmatic subnucleus of lightly labeled cells. Paraffin. *Scale*, 400 μm

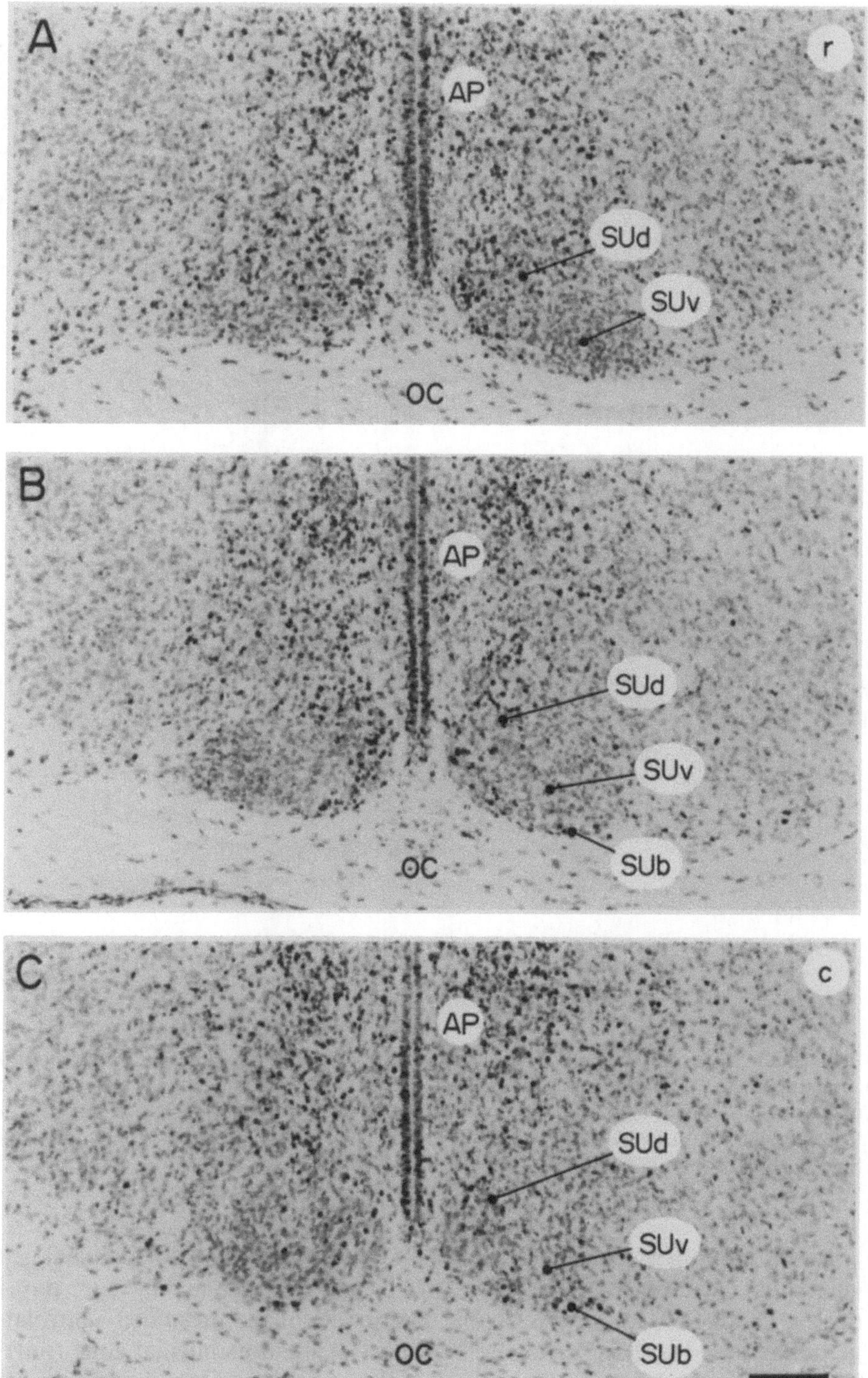

Fig. 105 A–C. Coronal thymidine radiograms, from rostral (**A**) to caudal (**B**), from a perinatal rat labeled on day E17. The neurons of the ventrolateral suprachiasmatic subnucleus are no longer labeled but some of dorsomedial subnucleus are still labeled. Labeled cells are also present in the basal suprachiasmatic subnucleus. Paraffin. *Scale*, 400 μm

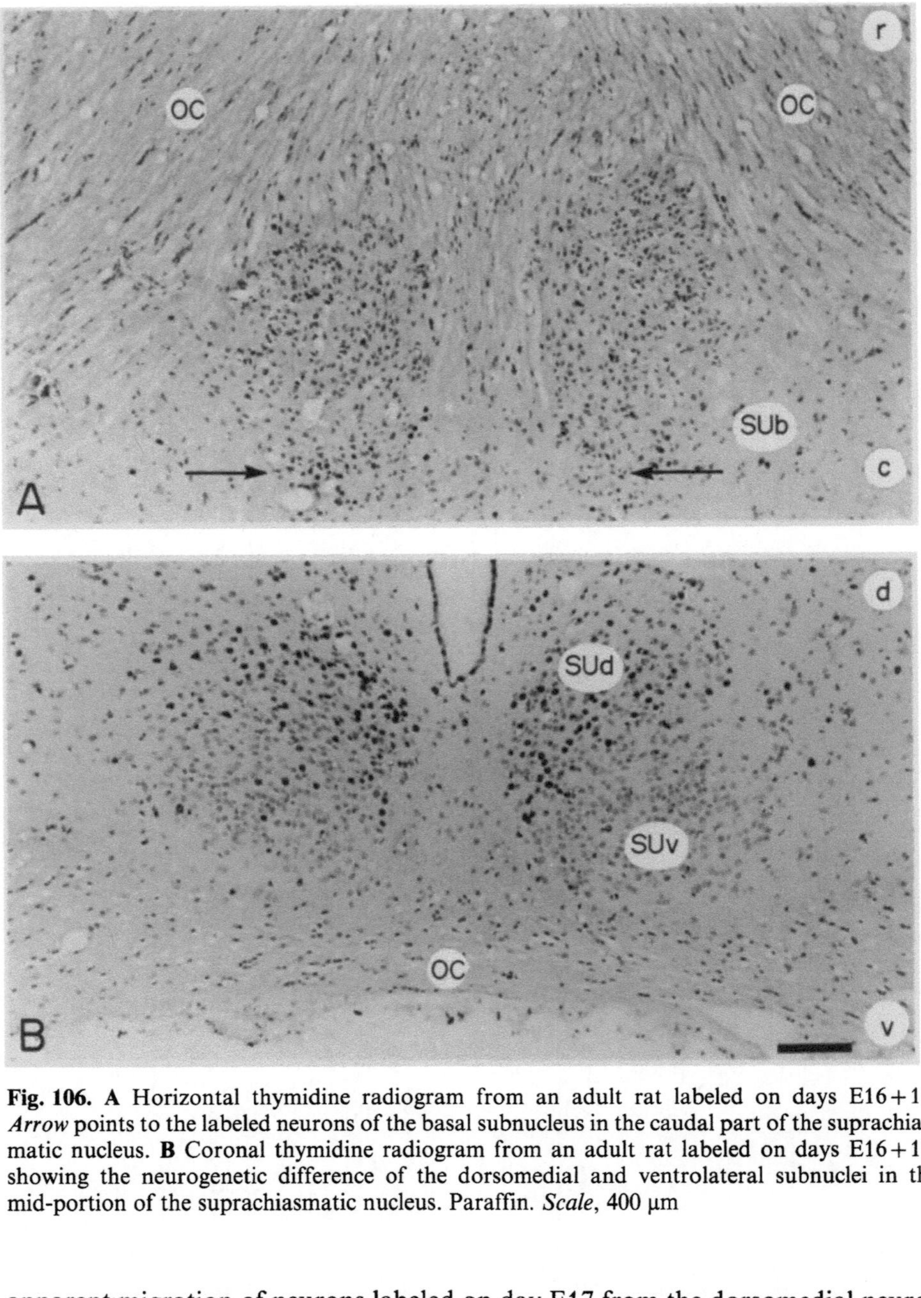

Fig. 106. A Horizontal thymidine radiogram from an adult rat labeled on days E16+17. *Arrow* points to the labeled neurons of the basal subnucleus in the caudal part of the suprachiasmatic nucleus. **B** Coronal thymidine radiogram from an adult rat labeled on days E16+17, showing the neurogenetic difference of the dorsomedial and ventrolateral subnuclei in the mid-portion of the suprachiasmatic nucleus. Paraffin. *Scale*, 400 μm

apparent migration of neurons labeled on day E17 from the dorsomedial neuroepithelium is illustrated in a rat injected on day E17 and killed on day E18 (arrow in Fig. 110A). In this animal many cells of the presumed ventrolateral germinal zone (suv) are unlabeled but most of the dorsomedial zone (sud) are labeled. The rapid settling of the late-generated neurons of the basal subnucleus is also indicated (SUb in Fig. 110).

Comments. The observations made in perinatal (day E22) rats confirm the chronology of the generation of suprachiasmatic neurons previously established in

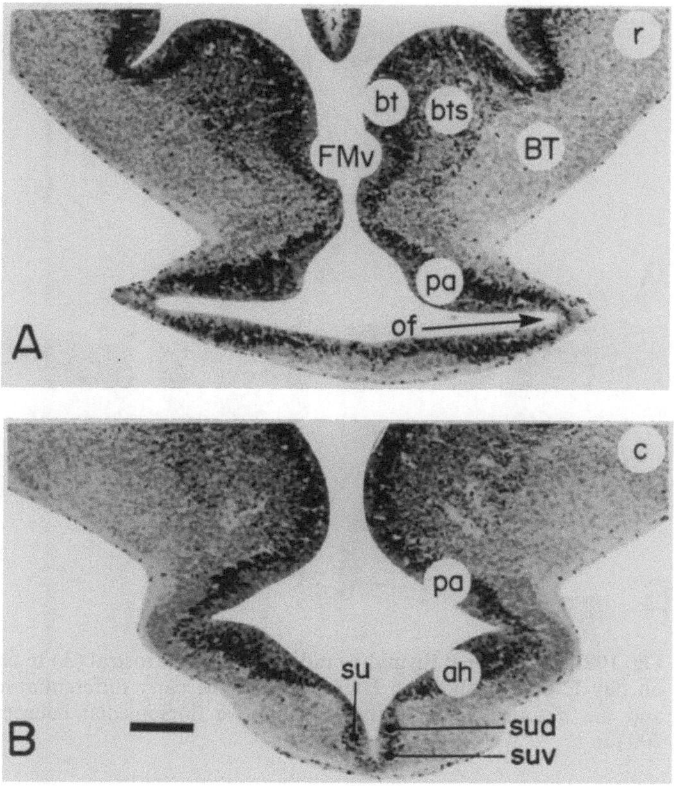

Fig. 107 A, B. Coronal thymidine radiograms, from rostral (**A**) to caudal (**B**), from a rat labeled on day E15 and killed 2 h later, showing the presumptive site of origin of suprachiasmatic neurons from an evagination of the anterior hypothalamic neuroepithelium. Paraffin. *Scale*, 200 μm

adult rats with long-survival thymidine radiography (Altman and Bayer 1978a: Fig. 5). But they modify our previous observations regarding neurogenetic gradients within the nucleus. As in our previous study, we have found a pronounced gradient from ventrolateral-to-dorsomedial but we could not confirm the gradient from rostral-to-caudal. The ventrolateral-to-dorsomedial gradient has been correlated by van den Pol (1980) with two cell groups of the suprachiasmatic nucleus, what we have called here the ventrolateral and dorsomedial subnuclei. The apparent rostral-to-caudal gradient we now attribute to the existence of a third subnucleus caudally, the basal subnucleus, largely composed of late-generated neurons.

Observations in sequential radiograms revealed the source of suprachiasmatic neurons in a transient ventromedian evagination of the neuroepithelium behind the optic fissure. This neuroepithelium shows high proliferative activity during the period of suprachiasmatic neurogenesis. The suprachiasmatic neuroepithelial zone is divisible into two parts, an earlier-declining ventral part (Fig. 110), which apparently produces the neurons of the ventrolateral subnucleus, and a later-declining dorsal component that may produce the neurons of the dorsomedial subnucleus and of the basal subnucleus. The locally proliferating cells seen in the differentiating suprachiasmatic nucleus on day E16 (Fig. 108)

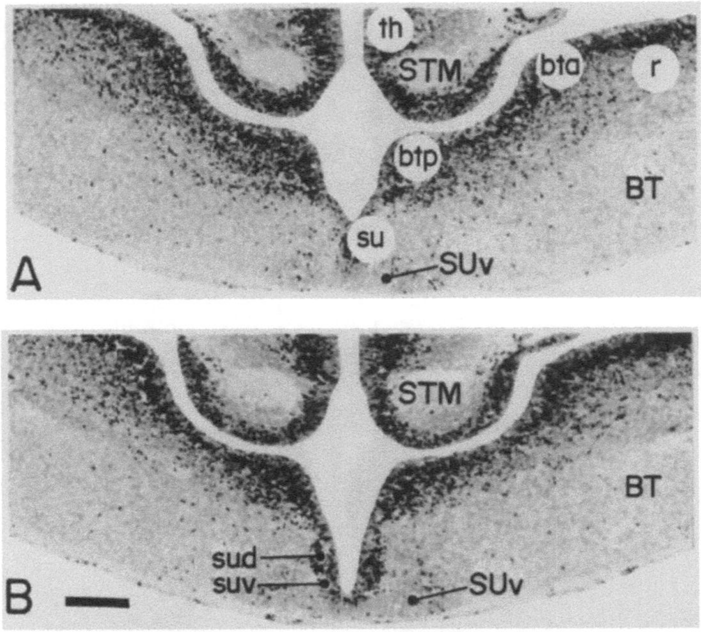

Fig. 108 A, B. Coronal thymidine radiograms, from rostral (**A**) to caudal (**B**) from a rat labeled on day E16 and killed 2 h later, showing the early differentiating ventrolateral subnucleus and the mitotically still active presumptive dorsomedial neuroepithelium. Paraffin. *Scale*, 200 μm

and, after cessation of suprachiasmatic neurogenesis on day E18 (Fig. 109B), may be the astroglia cells described by van den Pol (1980). The observations made in sequential radiograms clearly establish that the suprachiasmatic nucleus is not part of the preoptic area but is derived from the suprachiasmatic component of the anterior hypothalamic neuroepithelium.

The suprachiasmatic nucleus has been a subject of considerable physiological and behavioral research interest because of its role in the regulation of circadian rhythms. Bilateral destruction of the area of the suprachiasmatic nucleus abolishes the circadian rhythms of drinking and locomotion (Stephan and Zucker 1972; Moore and Eichler 1972), of the sleeping-waking cycle (Ibuka and Kawamura 1975), and of pituitary-adrenal functions (Moore and Eichler 1972). The effect on temperature regulation and reproductive functions is controversial (Rusak and Zucker 1979). It has been suggested that separate components of the suprachiasmatic nucleus may be involved in the regulation of pituitary-adrenal functions and body temperature (Abe et al. 1979). The evidence that the suprachiasmatic nucleus has at least two and possibly three structural divisions raises the possibility that they subserve different functions. The evidence reviewed above that the direct and indirect visual afferents distribute primarily in the ventrolateral subnucleus and that the vasopressin- and neurophysin-containing efferents originate in the dorsomedial subnucleus suggests the possibility that the former is the "sensory" component of the suprachiasmatic nucleus (perhaps mediating the photic entrainment of the circadian rhythm) and the latter its "effector" component (possibly the freely oscillating pacemaker). How-

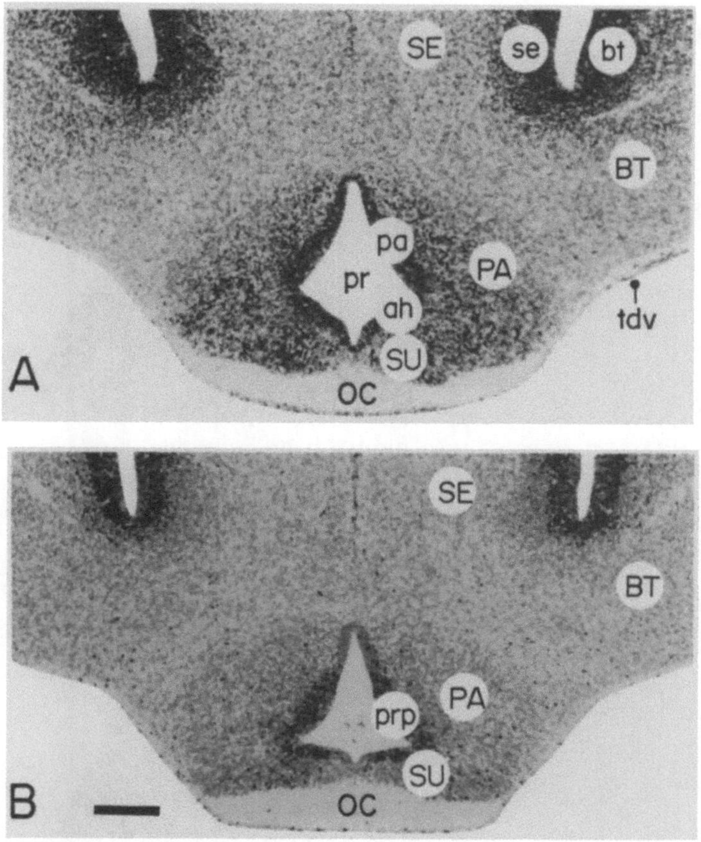

Fig. 109 A, B. Coronal thymidine radiograms from day-E18 rats labeled on day E16 (**A**) and on day E18 (**B**). In the latter, the labeled cells near the lumen are considered to be the source of preoptic periventricular neurons, and the locally multiplying cells within the suprachiasmatic nucleus may be astrocytes. Paraffin. *Scale*, 200 μm

ever, if this were the case, the observed neurogenetic pattern would be anomalous, because, unlike most brain regions, the neurons of the imputed receptor structure (the ventrolateral subnucleus) are generated before the imputed effector structure (the dorsomedial subnucleus).

7.2 The Arcuate Nucleus

Background. The arcuate (or infundibular) nucleus is an elongated structure in the mediobasal hypothalamus, extending all the way from the anterobasal nucleus to the mammillary body. It is composed of parvocellular neurons with small dendritic arbors (van den Pol and Cassidy 1982). The arcuate nucleus has received considerable attention recently because it appears to be a central component of the "parvicellular neurosecretory system" (Szentágothai 1964), or what others have called the tubero-hypophyseal (Björklund et al. 1973) or tuberoinfundibular (Lechan et al. 1982) system of the hypothalamus. It is conceptualized as a final common pathway for the regulation of anterior pituitary

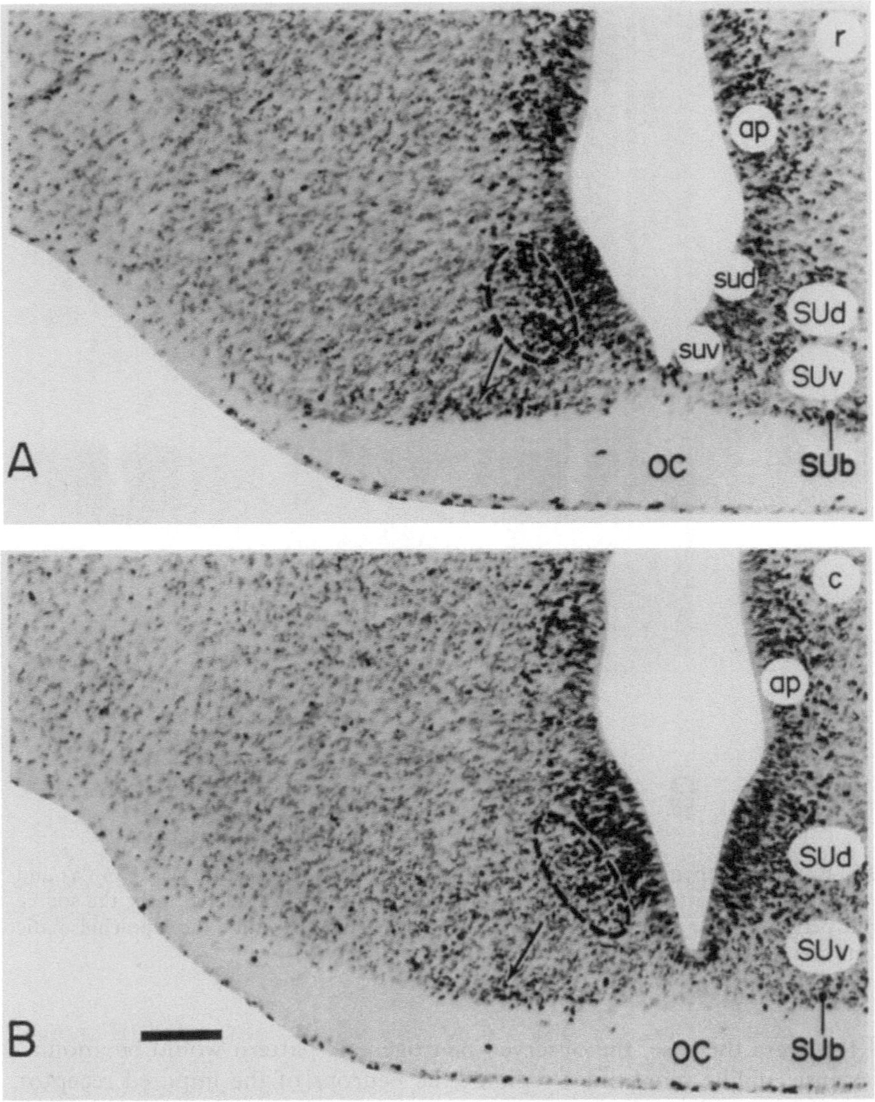

Fig. 110 A, B. Coronal thymidine radiograms, from rostral (**A**) to caudal (**B**), from a rat tagged on day E17 and killed on day E18. The migration of late-forming neurons from the dorsomedial suprachiasmatic neuroepithelium to the corresponding nucleus (*broken outline*) is indicated. The same active site may also be the source of the basal subnucleus neurons (*arrows*). Paraffin. *Scale*, 400 μm

functions by way of the tuberoinfundibular tract that terminates in the external layer of the median eminence and conveys releasing factors to the fenestrated capillaries of the pituitary portal system (Raisman 1972).

The arcuate nucleus is not the sole source of axons reaching the median eminence nor are its connections limited to the median eminence. Studies based on the retrograde transport of HRP and wheat germ agglutinin from the median eminence (Wiegand and Price 1980; Lechan et al. 1982) showed that within the hypothalamus the majority of the backfilled cells were in the arcuate nucleus, but some cells were also found in the periventricular area, parts of the paraventricular nucleus, and in the preoptic area. Among extrahypothalamic structures that send fibers to the median eminence are the diagonal band of Broca, the medial septum, and the brain stem (Lechan et al. 1982). At least a component of the arcuate fibers reaching the median eminence is rich in dopamine (Björklund et al. 1973; Nishizuka 1979). In addition to its efferents terminating in the median eminence, the arcuate nucleus also sends fibers to the ventromedial nucleus (Záborszky and Makara 1979), the paraventricular nucleus and preoptic area (Harris and Sanghera 1974; Joseph 1980), and several extrahypothalamic structures, including the amygdala, the septum, and the brain stem (Eskay et al. 1979; Joseph 1980).

Using immunocytochemical techniques, several releasing factors have been localized in arcuate neurons. A high concentration of luteinizing hormone-releasing hormone is present in some regions of the arcuate nucleus of the rat (Naik 1975; Baker et al. 1975; Kelly et al. 1982). It is relevant in this context that ultrastructural changes were observed in arcuate neurons of male rats after castration (Brawer 1971) and in female rats in relation to the different stages of the estrous cycle (King et al. 1974). Alpha-melanocyte-stimulating hormone cell bodies were found in highest concentration (Eskay et al. 1979), or exclusively (O'Donohue et al. 1979), in the arcuate nucleus. The immunoreactive fibers from the arcuate nucleus were found to distribute not only to the median eminence but to many hypothalamic structures and, in lower concentrations, to some extrahypothalamic regions (O'Donohue et al. 1979). Alpha-melanocyte-stimulating hormone in the rat has been implicated in the facilitation of learning of visual tasks (Sandman et al. 1972; Handelman et al. 1983). The arcuate nucleus has been found to be the site of the highest concentration of opiocortins (Knigge et al. 1981; Piekut and Knigge 1984). This family of related peptides includes, in addition to alpha-melanocyte-stimulating hormone, ACTH, beta-endorphin, and several other peptides. It has been claimed that in the rat arcuate nucleus these peptidergic neurons are located ventrally, while the dopaminergic neurons referred to earlier are localized dorsally (Bugnon et al. 1979). Finally, neurons immunoreactive to growth hormone-releasing factor are present in high concentration in the arcuate nucleus (Jacobowitz et al. 1983; Merchenthaler et al. 1984). Somatostatin, the hormone inhibiting the release of growth hormone, is either not present in the arcuate nucleus (Bennett-Clarke et al. 1980) or is found only in scattered cells (Dierickx and Vandesande 1979).

The development of the arcuate nucleus has not received much attention. According to Ifft (1972), the laterally situated arcuate neurons of the rat are produced on day E14 and those more medially situated on day E16. A similar lateral-to-medial gradient of neurogenesis was reported in the mouse (Shimada and Nakamura 1973). According to Koritsánszky (1979), the arcuate nucleus becomes delineated from its surroundings on day E18 and by this time a periventricular zone composed of small, fusiform cells can be distinguished from a lateral zone composed of larger, polygonal cells. In our earlier developmental study (Altman and Bayer 1978a) we have distinguished between the arcuate nucleus proper and an early-generated, cell-sparse, medially and subventricularly situated region which we have named the anterobasal nucleus (Sect. 6.2.1). Within the cell-dense arcuate nucleus proper our quantitative data indicated a rostral-to-caudal gradient. Anteriorly, peak production time was on day E15

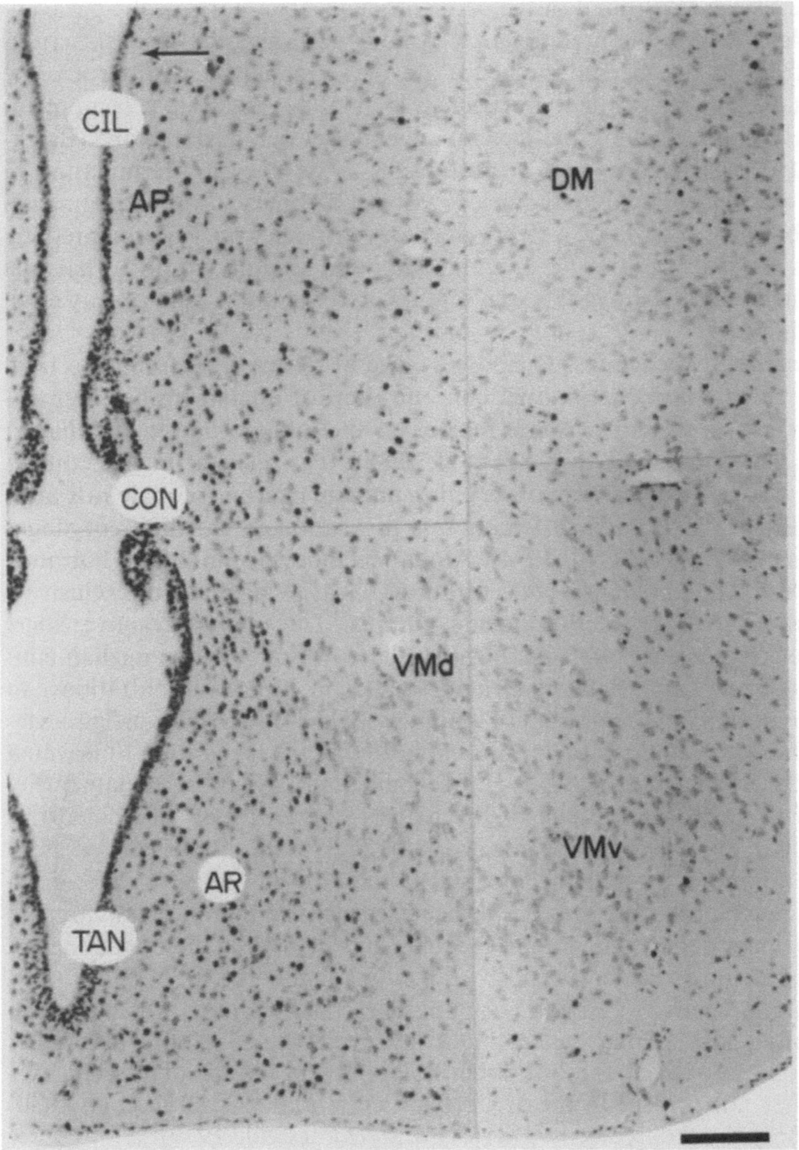

Fig. 111. Coronal thymidine radiogram from an adult rat labeled on days E16+17, showing the relatively late production of neurons of the arcuate nucleus and anterior periventricular area. Paraffin. *Scale*, 150 μm

and the last day of neurogenesis on day E17 (Altman and Bayer 1978a: Fig. 7). Posteriorly, peak production time was on day E18 and neurogenesis extended into day E19 (Altman and Bayer 1978a: Fig. 8), making this region the latest-generated neuronal population of the hypothalamus (Altman and Bayer 1978a: Fig. 23).

Results. The arcuate nucleus is distinguished from the other nuclei surrounding it laterally and dorsolaterally by the relatively late production of its neurons.

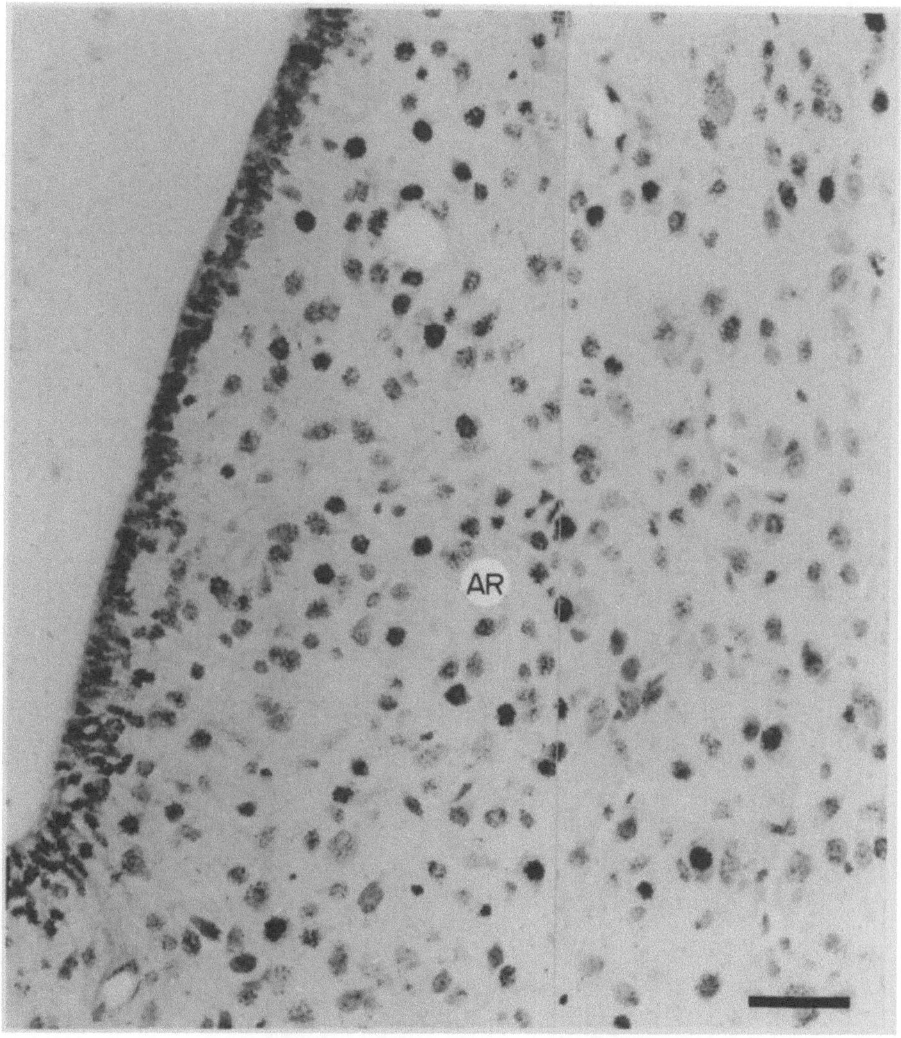

Fig. 112. Detail of Fig. 111. Note random distribution of heavily and lightly labeled cells throughout the arcuate nucleus. *Scale*, 50 μm

This is illustrated in an adult rat injected with [^{3}H]thymidine on days E16+17 (Fig. 111). At this rostral level most arcuate neurons are labeled in this injection group, some lightly, others heavily (Fig. 112). There is a distinct rostral-to-caudal gradient in the production of arcuate neurons. In adult rats injected on days E17+18, the arcuate neurons are mostly unlabeled rostrally (Fig. 113A). In the mid-anteroposterior portion of the arcuate nucleus there is an increase in the number of labeled cells (Fig. 113B), and caudally there is a high proportion of labeled cells (Fig. 113C). Some labeled cells are still scattered in the caudal arcuate nucleus in adult rats injected on days E19+20 (Fig. 114A) but are practically absent following later injections (Fig. 114B).

In rats injected on day E15 and killed on day E16, labeled cells form a fringe (arrows in Fig. 115) around the arcuate neuroepithelium (ar). The distance

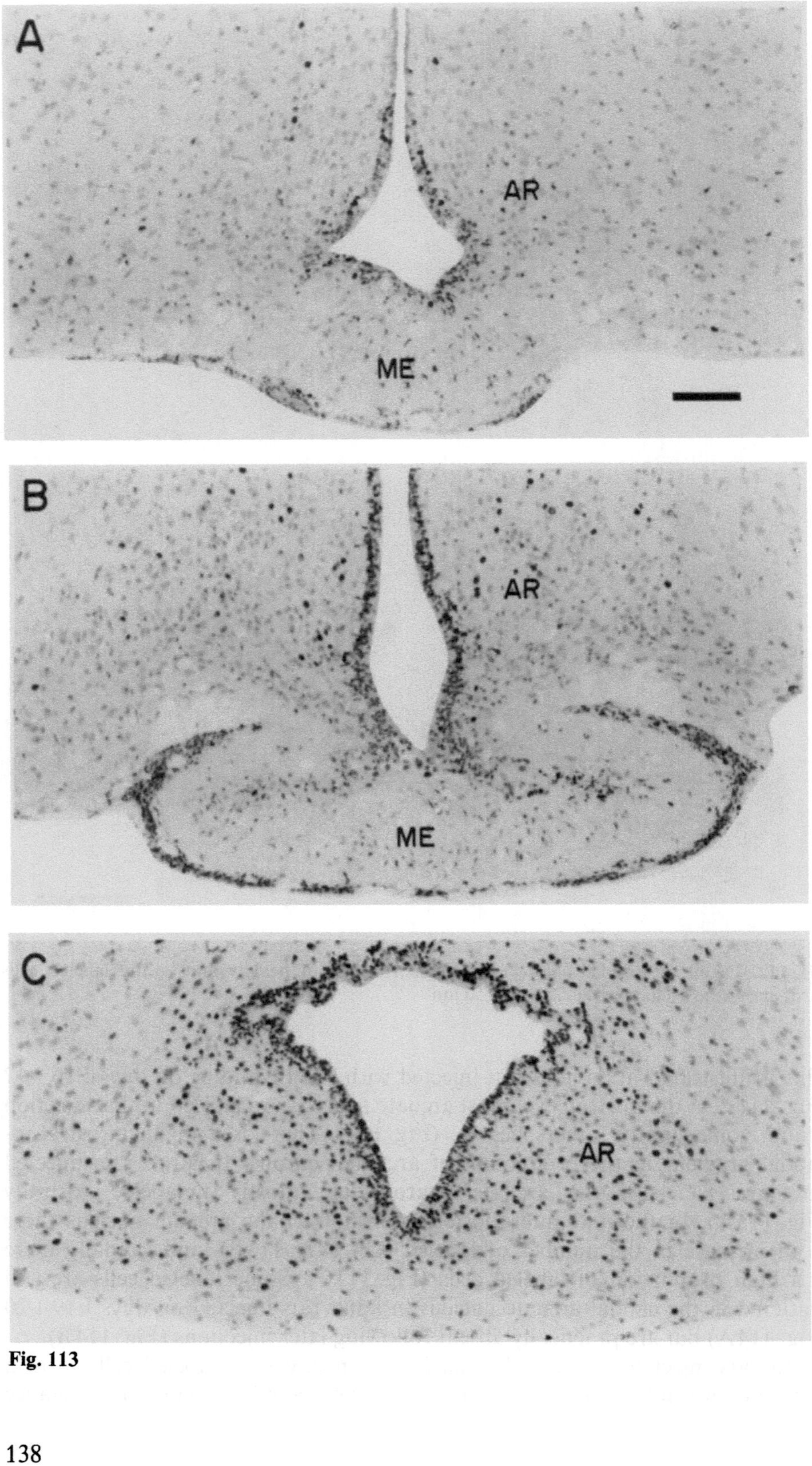

Fig. 113

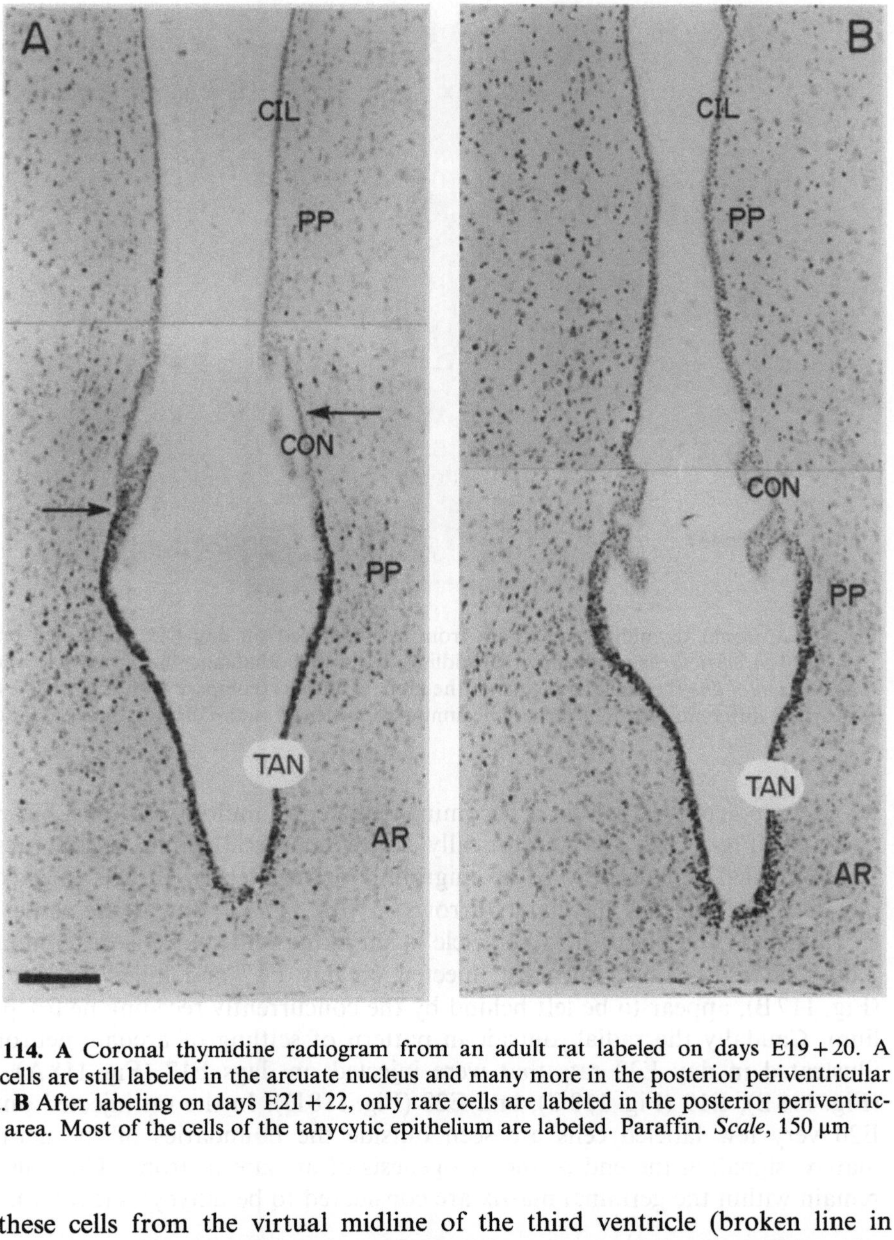

Fig. 114. A Coronal thymidine radiogram from an adult rat labeled on days E19+20. A few cells are still labeled in the arcuate nucleus and many more in the posterior periventricular area. **B** After labeling on days E21+22, only rare cells are labeled in the posterior periventricular area. Most of the cells of the tanycytic epithelium are labeled. Paraffin. *Scale*, 150 μm

of these cells from the virtual midline of the third ventricle (broken line in Fig. 115) suggests that these early-generated neurons occupy a position that will form the lateral border of the future arcuate nucleus (compare with Figs. 116C–117B). By day E17 the arcuate nucleus is recognizable laterally, as shown in radiograms of a rat injected on day E17 and killed 2 h later (Fig. 116), although its dorsal boundary is not clear. Rostrally (Fig. 116A),

◁ **Fig. 113A–C.** Coronal thymidine radiograms, from rostral (**A**) to caudal (**C**), from an adult rat labeled on days E17+18. Note the rostral-to-caudal neurogenetic gradient in the arcuate nucleus with mostly unlabeled cells rostrally and predominantly labeled cells caudally. Paraffin. *Scale*, 100 μm

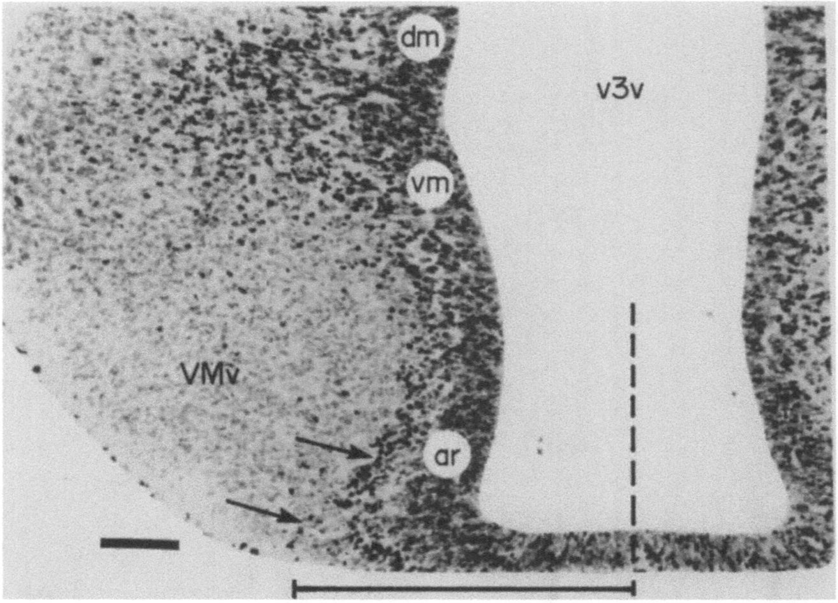

Fig. 115. Coronal thymidine radiogram from a rat labeled on day E15 and killed on day E16. *Broken vertical line* indicates the midline of the hypothalamic third ventricle and the *solid horizontal line* the far-lateral edge of the arcuate nucleus (compare with Fig. 117B) where the earliest-differentiating neurons are beginning to settle (*arrows*). Paraffin. *Scale*, 100 μm

the arcuate neuroepithelium is becoming depleted of multiplying cells, but more caudally (Fig. 116B, C) a mitotically active neuroepithelium is still present. These labeled cells appear to be migrating ventrolaterally in rats injected on day E17 and killed 1 day later (arrows in Fig. 117A). But at the same time the lumen of the basal third ventricle is shrinking and the cells differentiating 2 days later, as shown in a rat injected on day E17 and killed on day E19 (Fig. 117B), appear to be left behind by the concurrently receding neuroepithelium. Caudally the radial, outside-in pattern of settling of arcuate neurons is illustrated in day E22 rats that were injected on days E17 (Fig. 118A), E18 (Fig. 118B), E19 (Fig. 118C), and E20 (Fig. 118D). In the rat injected on day E20 very few labeled cells are seen outside the boundaries of the germinal matrix, signaling the end of the cytogenesis of arcuate neurons. The cells that remain within the germinal matrix are considered to be tanycytes (Sect. 8).

Comments. These results confirm the pronounced rostral-to-caudal gradient in the arcuate nucleus that we have found earlier in a quantitative study in adult rats. Rostrally, peak production of arcuate neurons occurs on day E15 and the last contingent of neurons are generated on day E17 (Altman and Bayer 1978a Fig. 7). Caudally, peak production is on day E18 and over 10% of the neurons are generated on day E19 (Altman and Bayer 1978a Fig. 8). This longitudinal gradient is coupled with a lateral-to-medial gradient, as previously described by Ifft (1972) in the rat, and by Shimada and Nakamura (1973) in the mouse. The lumen of the basal third ventricle does not shrink sufficiently to allow the settling of arcuate neurons until day E16 (Fig. 115) or E17

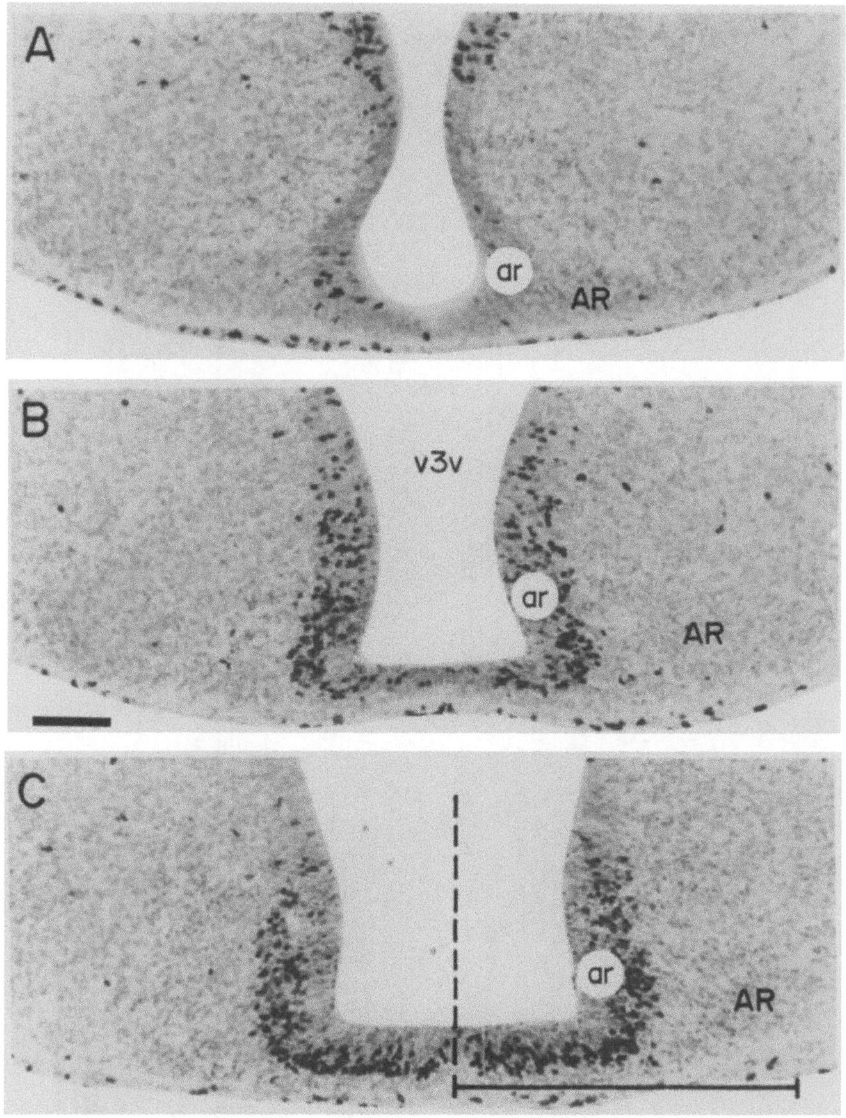

Fig. 116 A–C. Coronal thymidine radiograms, from rostral (**A**) to caudal (**C**), from a rat labeled on day E17 and killed 2 h later. Note the rostral-to-caudal gradient in the labeling of the putative arcuate neuroepithelium. Paraffin. *Scale*, 100 μm

(Fig. 116). This is compatible with the report of Halász et al. (1972) that arcuate neurons (presumably the earliest of them) already have synapses on them by day E16 and nerve terminals with synaptic vesicles are present in the median eminence by this time. According to Halász et al. (1972) and Kobayashi et al. (1968) nerve terminals in the median eminence of the rat increase greatly by day E18.

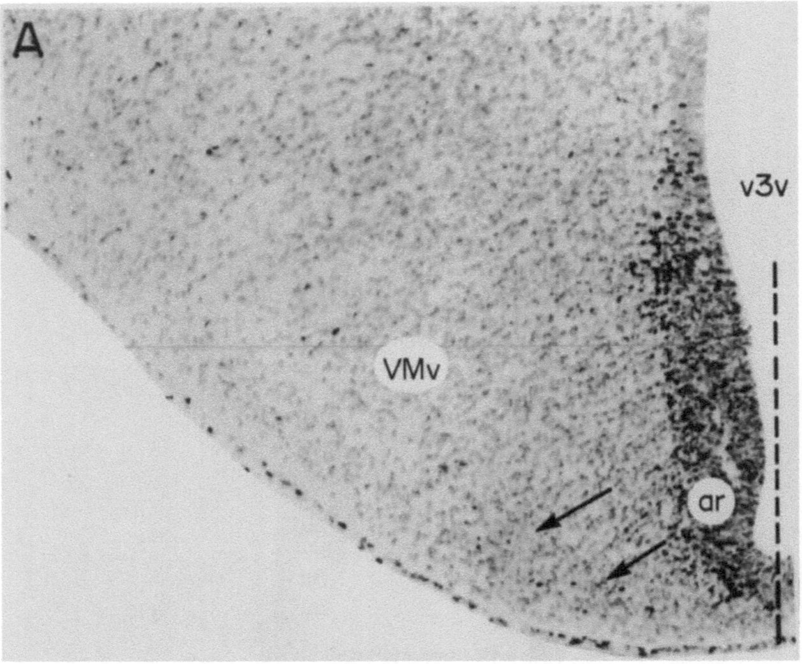

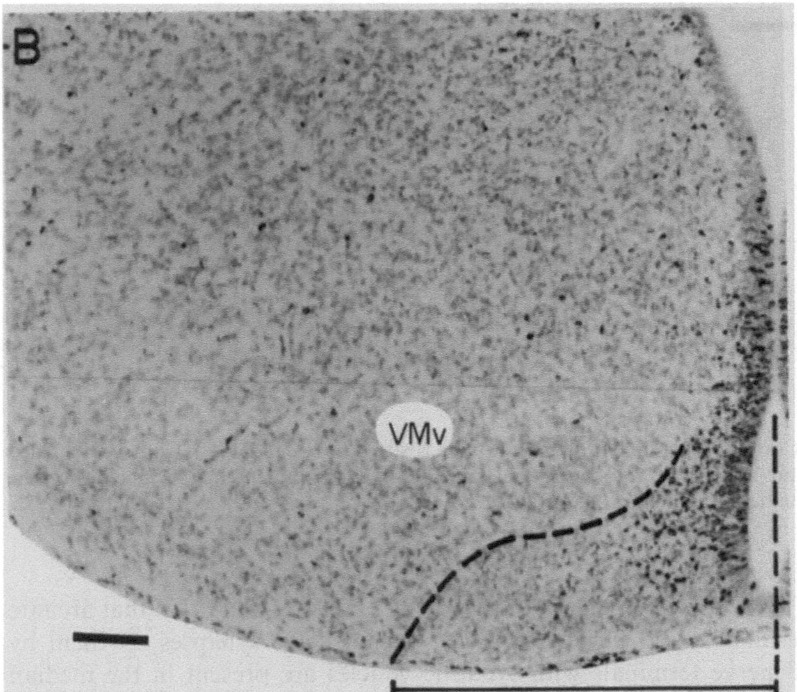

Fig. 117A, B. Coronal thymidine radiograms from rats tagged on day E17 and killed on day E18 (**A**) and E19 (**B**). Note the apparent migration of arcuate neurons ventrolaterally on day E18 (*arrows in* **A**). By day E19 the lumen of the ventricle has receded medially and the late-forming arcuate neurons appear to settle *in situ*, creating at this age a lateral-to-medial neurogenetic gradient. Paraffin. *Scale*, 100 μm

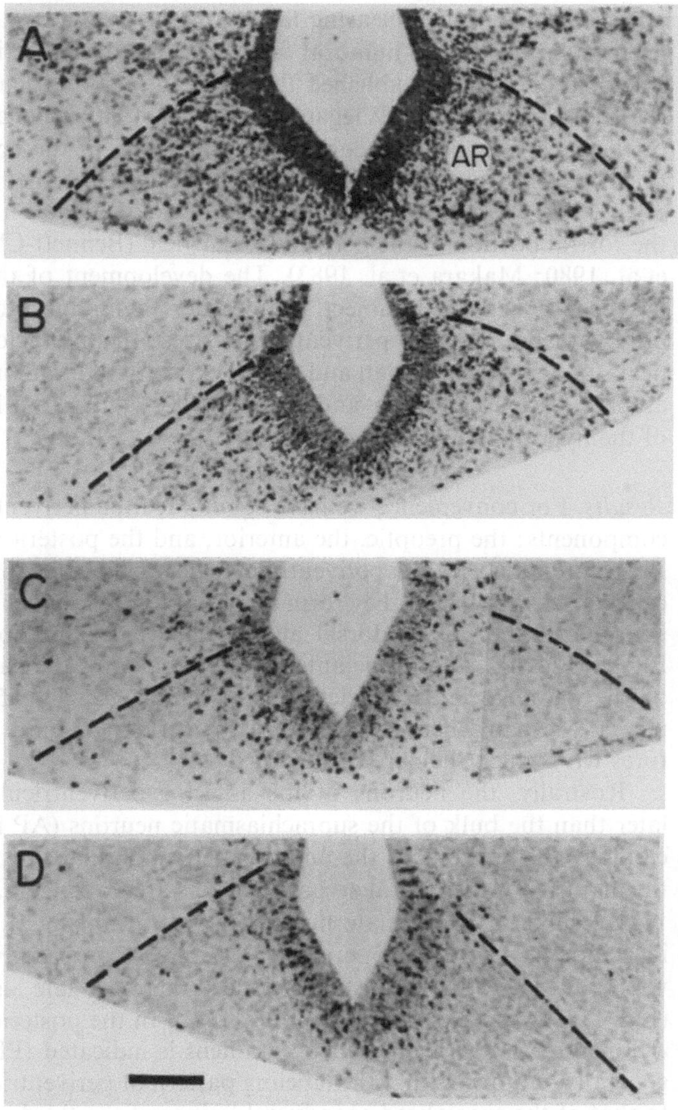

Fig. 118A–D. Coronal thymidine radiograms from perinatal rats labeled on day E17 (**A**), E18 (**B**), E19 (**C**), and E20 (**D**), showing the lateral-to-medial gradient in the generation of arcuate neurons. Paraffin. *Scale*, 100 µm

7.3 The Periventricular Field

Background. In addition to the two specialized nuclei of the midline hypothalamus, the suprachiasmatic nucleus and the arcuate nucleus, small neurons are scattered near the ependymal lining of the third ventricle from the preoptic area rostrally to the posterior hypothalamus caudally. These neurons appear to be a component of the "parvicellular neurosecretory system," associated with the "tuberohypophyseal tract" (Szentágothai 1964). This system was hypothesized to convey releasing factors to the anterior pituitary gland, and may itself be influenced by humoral factors in the third ventricle. Recent retrograde tracer studies have established that, indeed, periventricular neurons project to the median eminence (Wiegand and Price 1980; Lechan et al. 1982). A few of these periventricular neurons are embedded in the ependymal lining of the third ventricle and these, too, may project to the median eminence (Lechan et al. 1982). Among neuroactive peptides, somatostatin has been localized in the rostral portion of the periventricular field (Bennett-Clarke et al. 1980; Kahn et al. 1980; Makara et al. 1983). The development of the periventricular field has been a neglected subject. Our earlier thymidine radiographic study dealt only with the preoptic periventricular area. In this region, neuron production peaks on day E18 (Altman and Bayer 1978: Fig. 5), ranking it with the posterior arcuate nucleus (Altman and Bayer 1978: Fig. 8), the last-produced neurons of the hypothalamus.

Results. For convenience we have subdivided the periventricular field into three components: the preoptic, the anterior, and the posterior periventricular areas. In the preoptic area the periventricular field cannot be distinguished from adjacent nuclei with which they form a continuum. Examples are the periventricular neurons of the posterodorsal and posteroventral preoptic regions (PPD and PPV in Fig. 44), and in the anterior hypothalamus the parvocellular subnucleus of the paraventricular nucleus (PVp in Figs. 52, 53, 55–59). Typically, these medially situated neurons are labeled for a day or more longer than the structures with which they are contiguous.

Rostrally, the neurons of the anterior periventricular area are produced later than the bulk of the suprachiasmatic neurons (AP in Fig. 105), and more caudally the neurons of the anterior periventricular area are still labeled when the majority of the rostral arcuate neurons are no longer labeled (AP in Figs. 73, 111). Unlike in the arcuate nucleus, a rostral-to-caudal gradient could not be identified in the anterior periventricular area (compare Figs. 73 and 111). Periventricular neurons are still produced here as late as days E19 and E20 (Fig. 114A) and possibly later (Fig. 114B). In the posterior periventricular area a dorsal-to-ventral neurogenetic gradient is indicated (PP in Fig. 114A). More generally, it appears that the labeling pattern of periventricular neurons is linked to that of the specialized ependymal linings of the third ventricle. Thus in adult rats that were tagged on days E16+17, the cells of the ciliated and convoluted ependyma are still labeled (CIL and CON in Fig. 111) and so are also the adjacent dorsal periventricular neurons. In adult rats that were tagged on days E19+20, when only the cells of the tanycytic ependyma are labeled (TAN in Fig. 114A), labeled periventricular neurons are seen only in this ventral region. However, the production of tanycytic ependymal cells outlasts the production of periventricular neurons (Fig. 114B).

As midline elements, the neurons of the periventricular field migrate only a short distance from their sites of production. In Figs. 36 and 109B the site of production of preoptic periventricular neurons (prp) is indicated in rats injected on day E18 and killed 2 h later. The site of production of anterior periventricular neurons (ap) is indicated in Fig. 110. But because at these sites the production of the cells of the specialized lining of the hypothalamic third ventricle continues after the cessation of periventricular neurogenesis (Sect. 8), it is not possible to distinguish the precursors of periventricular neurons from the proliferative ependymal elements.

Comments. The dispersed periventricular neurons share certain properties with the compactly packed neurons of the suprachiasmatic nucleus and, more particularly, the arcuate nucleus. Among these are their size, midline location, presumed control of the anterior pituitary gland, and, last but not least, the late production of their neurons. The classification of the periventricular field with the suprachiasmatic nucleus and the arcuate nucleus, as components of the midline hypothalamus and products of the "third wave" of hypothalamic neurogenesis, seems justified. The fact that the production of periventricular neurons slightly outlasts other components of the midline hypothalamus, and that there is a chronological link between the regional production of periventricular neurons and the cells of the specialized ependyma, suggests an intimate association between these two elements of the juxtaventricular hypothalamus.

8 Development of the Specialized Linings of the Third Ventricle

Background. While in many regions of the hypothalamus the wall of the third ventricle is composed of typical ependymal cells (cuboid in shape and ciliated on the side facing the lumen), in some other regions, especially in the posteroventral region of the anterior hypothalamus and the ventral region of the posterior hypothalamus, the ventricle is lined by specialized cells (Bruni et al. 1972; Kozlowski et al. 1973). Among these specialized cells, the best known are the tanycytes that line the ventricle above and around the arcuate nucleus (Colmant 1967; Bleier 1971; Millhouse 1971). The specialized ventricular cells have been implicated in secretory activity (Leonhardt 1966; Defendini and Zimmerman 1978) and in the uptake and transport of hormones (Kobayashi et al. 1972; Scott et al. 1974; Joseph and Knigge 1978). In our previous study in mature rats (Altman and Bayer 1978c) we have distinguished three types of specialized linings in the third ventricle: the convoluted ependyma, the laminated epithelium, and the tanycytic epithelium. Our study of long-survival thymidine radiograms indicated that the cells of the convoluted ependyma near the paraventricular nucleus and adjacent regions are produced relatively early, while the cells of the laminated epithelium near the dorsomedial and ventromedial nuclei, and the cells of the tanycytic epithelium near the arcuate nucleus, are produced later, mostly peri- and postnatally. We have suggested that the cells of these specialized ventricular linings arise from the same germinal matrix as the neurons of the contiguous hypothalamic nuclei with which they may be structurally and functionally associated. If so, they would represent the "fourth wave" of hypothalamic cells produced; though not truly a wave because these cells do not migrate.

Results. In adult rats injected on days E16+17 (Fig. 111), practically all the cells of the ciliated ependyma (CIL), convoluted ependyma (CON), and tanycytic epithelium are labeled, as are the neighboring neurons of the anterior periventricular area (AP) and arcuate nucleus (AR). In adult rats injected on days E17+18 (not shown), many of the cells of the ciliated and convoluted ependyma are no longer labeled but those of the laminated and tanycytic epithelia are labeled. In adult rats injected on days E19+20, most cells of the ciliated ependyma are no longer labeled (Fig. 114A). In this injection group, the cells of the convoluted ependyma are not labeled dorsally (Fig. 119), but some are still labeled ventrally (Figs. 114A, 119). All the cells of the laminated epithelium (LAM in Fig. 119) and the tanycytic epithelium (Figs. 114A, 119) are labeled. In adult rats injected on days E21+22 (Fig. 114B), the convoluted ependyma is altogether free of labeled cells but the tanycytes are labeled. As we have shown earlier (Altman and Bayer 1978c: Figs. 8–10), some of the cells of the

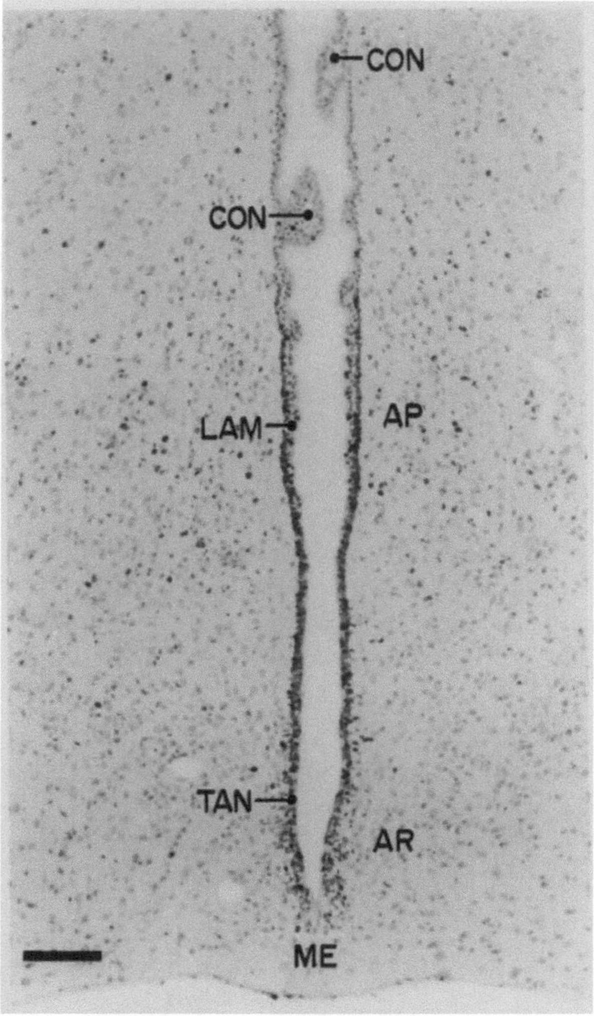

Fig. 119. Coronal thymidine radiogram from an adult rat labeled on days E19 + 20. Paraffin. *Scale*, 100 μm

laminated and tanycytic epithelium are still labeled in rats injected during the 1st and 2nd week of postnatal life.

In rats injected on day E18 and killed 2 h later (arrow in Fig. 75D), labeled cells persist in the "neuroepithelium" which on previous days was presumed to produce the neurons of the dorsomedial nucleus. Since dorsomedial neuron production comes to an end on day E17, we have suggested earlier (Sect. 6.3.1) that these multiplying cells now commence to produce cells for the laminated epithelium. Cell proliferation apparently increases in this region, as well as more ventrally and caudally, on the subsequent days. In rats injected on day E20 and killed 2 h later, labeled cells are present in the lining of the ventricle at the levels of the dorsomedial and ventromedial nuclei (Fig. 120A). The dorsal component of this proliferative region is designated as the precursors of the laminated epithelium (lam), and its ventral component as the germinal cells

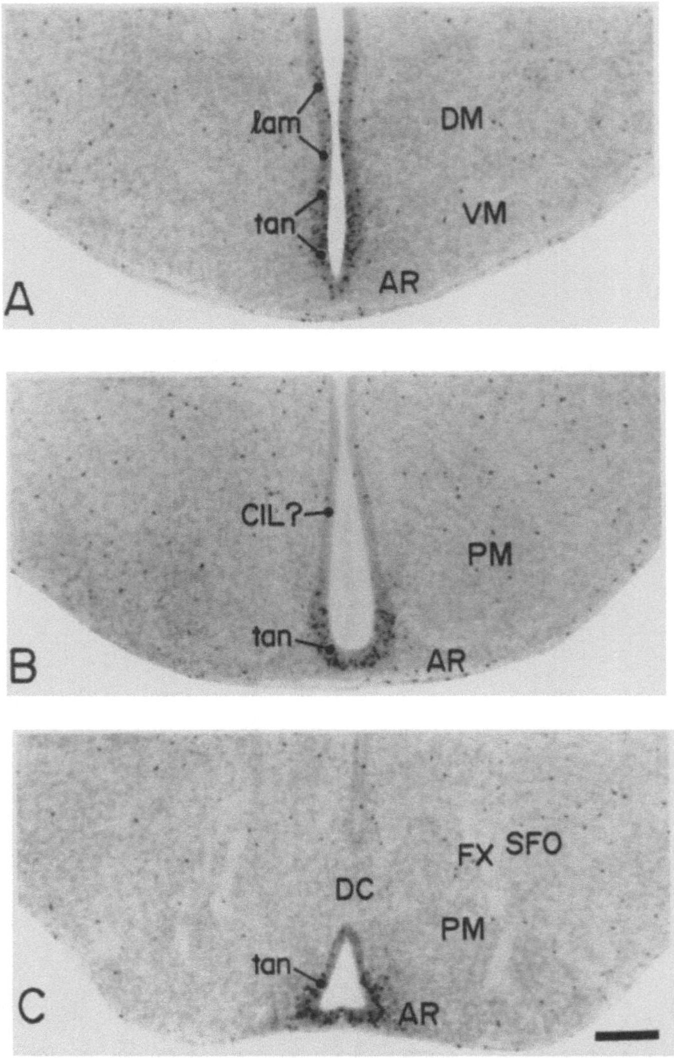

Fig. 120 A–C. Coronal thymidine radiograms, from rostral (**A**) to caudal (**C**), to show the persistence of cell labeling at the future sites of the laminated and tanycytic epithelium. Paraffin. *Scale*, 200 μm

of the tanycytic epithelium (tan). More caudally, the ependymal cells in the vicinity of the premammillary nuclei appear to have differentiated; these may be ciliated ependymal cells (CIL? in Fig. 120B). Ventrally at this level, and more caudally (Fig. 120C), the ventricular zone adjacent to the arcuate nucleus is composed of labeled cells. These are designated as the tanycyte precursors (tan) of the arcuate nucleus. The pattern persists on the subsequent days.

Comments. After cessation of neurogenesis at a particular site, the proliferative precursors of neurons (or possibly cells of another, dormant cell line) commence to produce ependymal cells. A subclass of these ventricular elements in some hypothalamic regions are those that form specialized linings of the third ventricle. We have previously identified and illustrated these (Altman and Bayer 1978c) as the convoluted ependyma, the laminated (multi-layered) epithelium, and the tanycytic epithelium. The convoluted ependyma is contiguous with the paraventricular nucleus and adjacent regions; the laminated epithelium appears to be related to the dorsomedial nucleus; and the tanycytic epithelium may have two parts, one related to the ventromedial nucleus and the other to the arcuate nucleus. With the exceptions of the convoluted ependyma, the production of these specialized ventricular cells continues through the period of infancy. The postnatal production of these elements and their maturation have been described previously (Altman and Bayer 1978c).

9 Development of Some Glial Components of the Hypothalamus

Background. In young embryos killed 2 h after the administration of [^{3}H]thymidine, most of the labeled proliferating cells are concentrated in the neuroepithelium lining the third ventricle. Some scattered labeled cells are also seen within the differentiating tissue of the hypothalamus but these represent but a minute fraction of the total number of labeled cells. (See the short survival radiograms from day E15 rats in Figs. 9–13, from day E16 rats in Fig. 108, and from day E17 rats in Figs. 14–16.) But as development proceeds and cell labeling diminishes and then ceases in the ventricular lining, there is a relative increase in the scattered labeled cells in the tissue of the hypothalamus (compare the short survival radiograms from a day E20 rat in Fig. 120). In our previous study of the development of the rat spinal cord (Altman and Bayer 1984) we proposed that these cells multiplying outside the neuroepithelium after the cessation of neurogenesis are precursors of glia cells.

Results. In rats injected with [^{3}H]thymidine on day E17 and killed 2 h later (Fig. 121 A), some scattered cells are labeled in the hypothalamus outside the

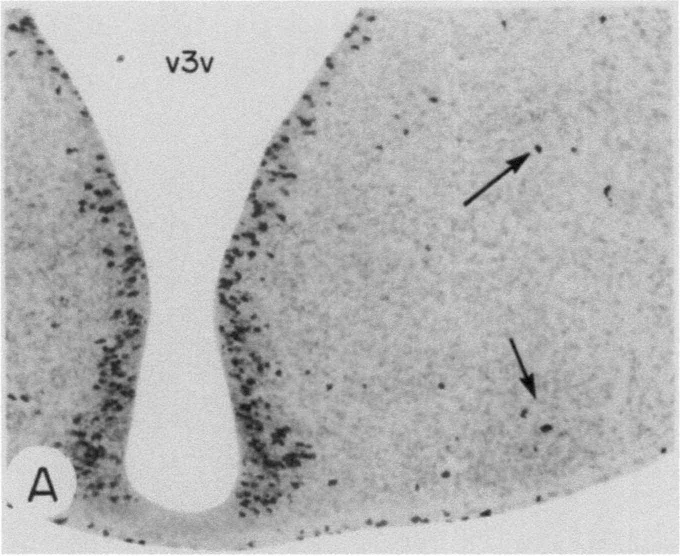

Figs. 121 and 122. Coronal thymidine radiograms from rats killed 2 h after injection on days E17 (Fig. 121 A), E18 (Fig. 121 B), E19 (Fig. 122 A), and E20 (Fig. 122 B). Note the apparent increase in the proportion of locally multiplying cells. Paraffin. *Scale*, 100 μm

proliferative neuroepithelium. The pattern remains similar in short-survival radiograms from rats tagged on day E18 (Fig. 121 B). There may be some increase in the locally multiplying cells in rats injected on day E19 (Fig. 122A). By day E20 (Fig. 122B), when ventricular cell proliferation has come to an end in all regions except where the laminated and tanycytic epithelia are formed (Fig. 120), the concentration of locally multiplying cells has greatly increased (Fig. 122B). Local cell proliferation remains high in rats labeled with [^{3}H]thymidine on day E21 (Fig. 123). While the distribution of these locally multiplying cells is not uniform, systematic regional differences within the hypothalamus could not be discerned.

Comments. We have not attempted to quantify here, in the context of the embryonic development of the rat hypothalamus, the time course of local cell proliferation that produces glia cells, because it is known that gliogenesis in the rat extends far into the postnatal period (e.g., Altman 1966). What the present observations establish is that there is a low rate of extraventricular cell proliferation at early embryonic stages, and that this is followed by a spurt on day E20, 1–2 days after neurogenesis has come to an end. It is interesting in this context that in the spinal cord, where neurogenesis ceases earlier, this spurt occurs on day E18 (Altman and Bayer 1984: Fig. 120B). Further work is needed to establish the exact identity and fate of these locally multiplying nonneuronal elements. It is conceivable that the early-multiplying and late-multiplying precursors produce different types of satellite cells. Another question that remains to be resolved is whether or not the precursors of these locally multiplying cells ultimately derive from the ventricular germinal matrix and, if so, when do they become separated from it.

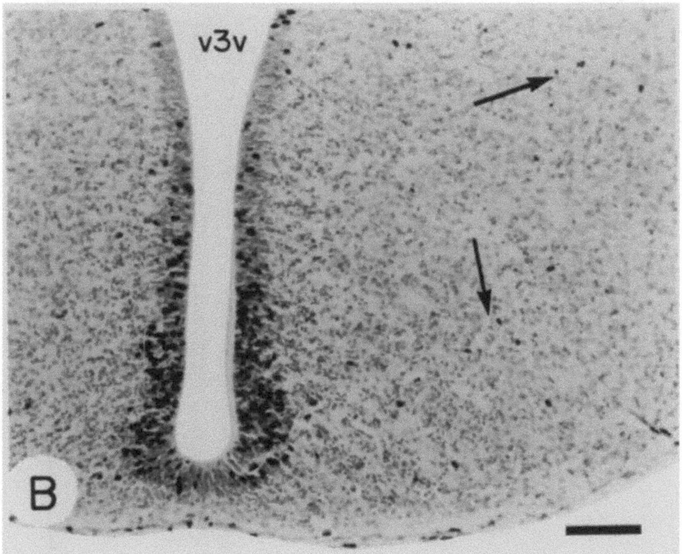

Fig. 121 B

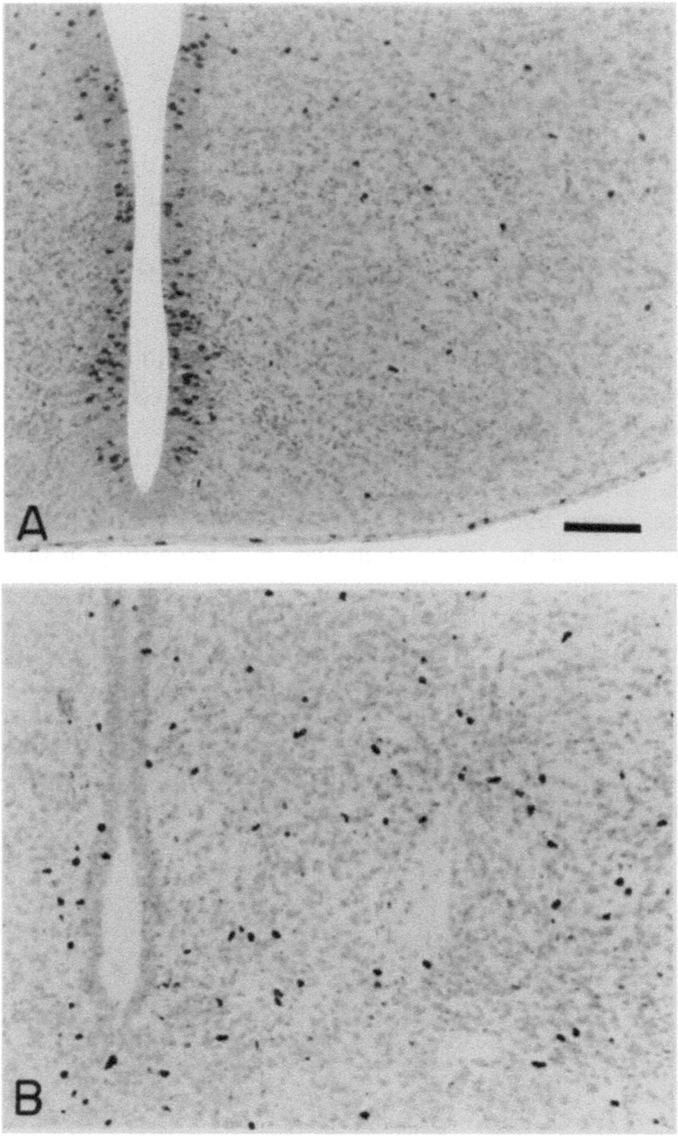

Fig. 122. Legend see Fig. 121

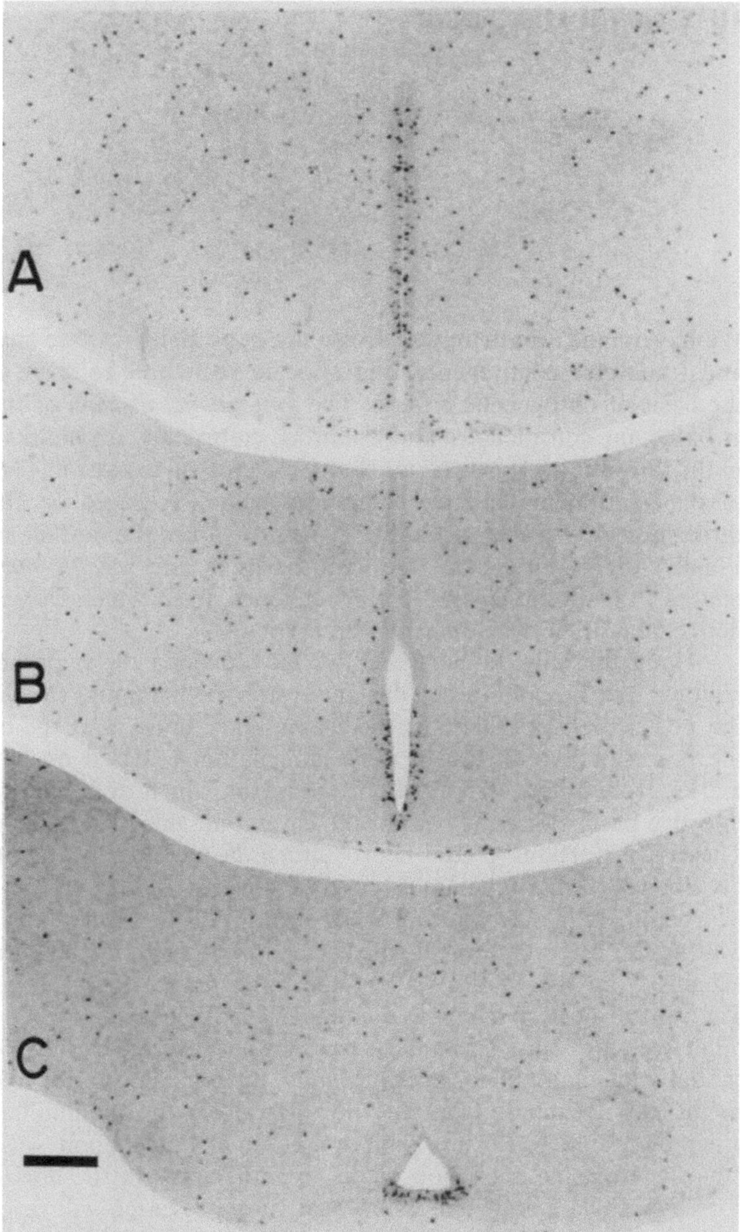

Fig. 123 A–C. Coronal thymidine radiograms, from rostral (**A**) to caudal (**C**), from a rat labeled on day E21 and killed 2 h later. Regional differences in the distribution of locally multiplying cells are not discernible. Paraffin. *Scale*, 200 µm

10 General Discussion

We have made an attempt to divide the hypothalamus into major longitudinal and transverse components, and allocate structures to these components, on the basis of ontogenetic criteria. The *longitudinal division* of the hypothalamus is based on conspicuous and enduring embryonic landmarks along the base of the third ventricle neuroepithelium from rostral to caudal. The *vertical division* of the hypothalamus, from lateral to medial, is based on the distinction of three successive waves of migratory neurons from the midline neuroepithelium. Finally, the *subdivision* of the hypothalamus *into specific nuclei* and *areas* is related to identifiable neuroepithelial zones, from which they are presumed to derive, and other developmental criteria.

Along the longitudinal plane we have divided the hypothalamus into five regions: the preoptic area, the anterior hypothalamus, the ventral region of the posterior hypothalamus, the mammillary body, and the caudal region of the posterior hypothalamus. This parcellation is based on conspicuous landmarks, from anteroventral to posterodorsal, which divide this germinal matrix into five parts along the base of the embryonic third ventricle neuroepithelium. These five portions of the neuroepithelium are named after the longitudinal divisions of the hypothalamus that they give rise to (Figs. 3, 4, 7, 8). The preoptic neuroepithelium extends from the rostral tip of the third ventricle to the rostral wall of the optic fissure caudally. The anterior hypothalamic neuroepithelium extends from the caudal wall of the optic fissure to the infundibular recess. The ventral posterior hypothalamic neuroepithelium extends from the infundibular recess to the mammillary recess. The mammillary neuroepithelium caps caudally the mammillary recess. Finally, the dorsal posterior hypothalamic neuroepithelium extends from the dorsal wall of the mammillary recess to the inner diencephalic flexure. This longitudinal parcellation was applied to the most heterogeneous portion of the hypothalamus in the horizontal plane, the core hypothalamus. It is possible that future research will reveal some unitary features shared by structural components of each of these divisions, and some differential features that distinguish components of different divisions, but such are not presently known. For instance, we cannot assign a unitary function or close association within the rostral region of the anterior hypothalamus to the anterior hypothalamic nuclei, on the one hand, and the paraventricular and supraoptic nuclei, on the other. Conversely, a unitary function (the control of reproductive behavior) has been attributed to components of the "preoptic-anterior hypothalamic continuum," that is, structures located in separate divisions of the hypothalamus. Such considerations suggest that the longitudinal division of the hypothalamus based on fissures and recesses in the neuroepithelium does not reflect a significant structural or functional discontinuity. Therefore, the longitudinal division of the hypothalamus is considered to be one of convenience, with the advantage over previous parcellations that it is based on clearly delineable embryonic landmarks.

However, the second division of the hypothalamus into three tiers from lateral to medial, the reticular, the core and midline, may have functional significance. This parcellation is based on the temporal order of the production and migration of neurons from the third ventricle neuroepithelium in three successive waves (Fig. 17). The *reticular hypothalamus* is composed of structures populated by neurons of the first wave that leave the neuroepithelium while it is still in a far-lateral position around the bulging lumen of the third ventricle (Fig. 17A). We included in the reticular hypothalamus three structures, the lateral hypothalamus, the entopeduncular nucleus, and the zona incerta. As we have summarized it in the relevant sections (5.1, 5.2, 5.3), these three structures share a morphological feature in that their diffusely distributed cell bodies are embedded in and related to a heterogeneous collection of ascending and descending fiber tracts. They may also share a functional trait in that they have been implicated in an as yet vaguely defined nonspecific category of sensory and, particularly, motor processes, what classical neurologists have referred to as extrapyramidal functions. Although we have used the term "reticular" in the anatomical sense, as a field composed of diffuse cell bodies in a matrix of fibers, the functional implication of the term, in the sense of a nonspecific activating mechanism (Magoun 1950), cannot be ruled out in this context. We have reviewed earlier (Sect. 5.1.1) the evidence for the involvement of the lateral hypothalamus in a variety of functions (aggression, feeding, attention, activity level) that can be thought to represent the motivational aspects of behavior. The lateral hypothalamus thus differs from other components of the hypothalamus in that it is not involved in the regulation of autonomic and hormonal functions but, rather, in skeletal-motor activities. Thus, the relationship between the lateral hypothalamus and the other proposed components of the reticular hypothalamus, the zona incerta and the entopeduncular nucleus, may be a closer one than hitherto assumed.

We shall leave aside for a moment the core hypothalamus and consider the *midline hypothalamus,* composed of the third wave of neurons of the hypothalamic neuroepithelium. The structures belonging to this system are the expanded periventricular field in the vicinity of the third ventricle and two specialized structures, the suprachiasmatic nucleus and the arcuate nucleus. As we have noted earlier (Sect. 7), the midline hypothalamic structures have many shared characteristics: they are composed of small neurons, many of these neurons contain releasing hormones or related peptides, and, with the exception of the suprachiasmatic nucleus, they are sending efferents to the median eminence. This system of late-produced neurons may thus constitute the "hypophysiotropic" area of the hypothalamus, representing the final common pathway of hypothalamic regulation of the anterior pituitary gland.

The *core hypothalamus* is the most heterogeneous part of the hypothalamus. It is composed of the second wave of neurons generated by the hypothalamic neuroepithelium. It was with reference to this system that we have used the longitudinal division of the hypothalamus into five major components. This system, with the exception of the supraoptic and paraventricular nuclei (which are the final common pathways to the posterior pituitary gland), may also

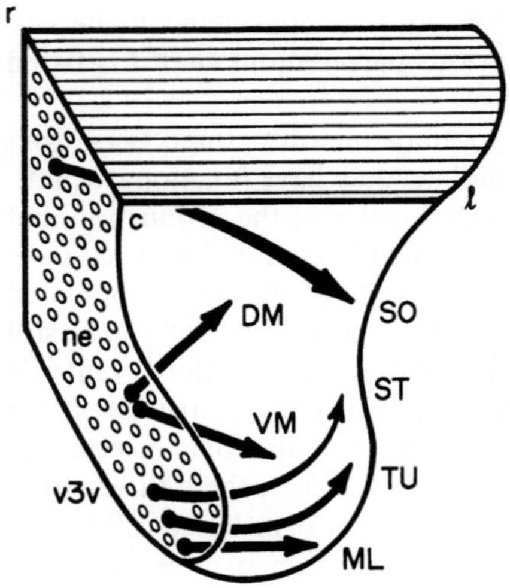

Fig. 124. Schematic illustration of migratory trajectories that deviate from the simple radial pattern

be conceived to have a unitary function in the weak sense that its nuclei constitute a "bridge" between the hypophysiotropic hypothalamus and the rest of the central nervous system. But it may be more useful to stress the structural and functional heterogeneity of the core hypothalamus which is relatable to the configurational heterogeneity of the neuroepithelium at the time that its neurons are generated.

The configurational complexity of the wall of the hypothalamic neuroepithelium is not due to large fissures and recesses but to less conspicuous or enduring evaginations, invaginations, thickenings, and thinnings of the germinal matrix, producing an undulating vertical surface that changes dynamically over time. With the combined use of short-survival, sequential, and long-survival thymidine radiography we have attempted to identify the concave and convex segments of the neuroepithelium, and its thickening and thinning patches, with discrete hypothalamic structures for which they are producing neurons. The time of origin of neurons of a particular nucleus, as determined by long-survival radiography, was the clue as to when during embryonic development its mitotically active neuroepithelial zone should be sought in short-survival radiograms. The examination of sequential radiograms made it possible to trace the migration of neurons from their presumed production site to their settling areas. This approach allowed us to assign to the hypothalamus structures that have not been considered to be part of it (e.g., the subthalamic nucleus) and exclude another (the bed nucleus of the stria terminalis). The migratory routes taken by neurons were often radial but we have encountered several modified trajectories (Fig. 124).

The implicaton of this approach is that the structural and functional heterogeneity of the core hypothalamus is due to a corresponding heterogeneity of the hypothalamic neuroepithelium in which the proliferating cells are determined to produce specific kinds of neurons before the onset of cell differentiation (cessation of cell multiplication). If this is correct, then there is in the hypotha-

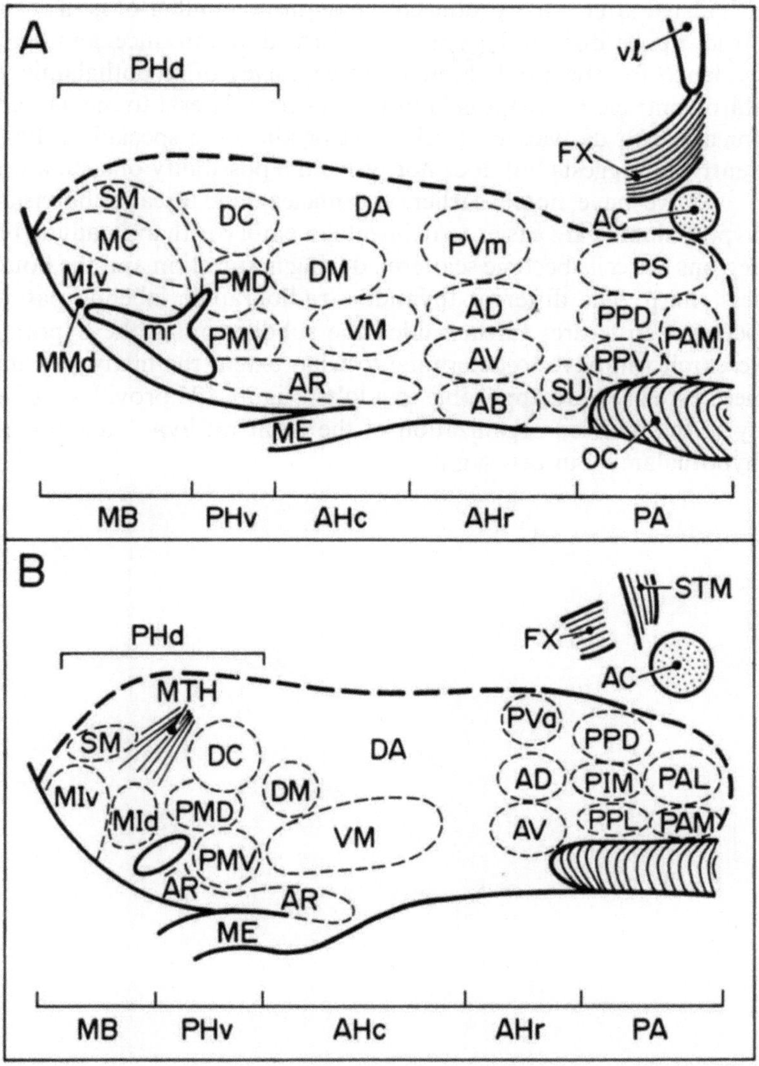

Fig. 125 A, B. Schematic illustration of the inferred organization of the adult rat hypothalamus in sagittal sections from medial (**A**) to lateral (**B**). The pattern is based on observations in late fetal not adult rats

lamic neuroepithelium a "preformed" mosaic of the discrete structures that it is programmed to produce. Evidently, the task of mapping of the third ventricle neuroepithelium requires much further work. Future neurochemical research may examine to what extent the chemical properties a specific class of neurons are already expressed by the proliferating cells of its putative neuroepithelial zone. Quantitative studies may be attempted to determine whether or not the putative neuroepithelial zone of a nucleus produces cells in sufficient numbers to account for the total population of neurons of that structure. The present evidence leaves unanswered many related questions. Are the discrete cell lines sharply separated from each other at boundary regions? Does a cell line become exhausted after it has produced the requisite number of neurons for a particular structure or does it become transformed to produce another cell line? The evidence for the production of three waves of hypothalamic neurons as the third ventricle neuroepithelium recedes from lateral to medial, and the evidence for a fourth derivative of cells that produce the specialized lining of the third ventricle, suggests but does not prove the possibility of such a transformation.

As we have noted earlier, the nuclear and areal subdivisions of the core hypothalamus are easier to delineate in embryos than in adults because in many regions the cells become scattered during maturation and the boundaries become less sharp. The different thymidine radiographic labeling patterns within and between structures further aided the subdivision of the hypothalamus. Future research will have to determine to what extent the many new nuclear divisions here proposed are applicable to adults. Figure 125 provides a schematic summary of the inferred organization of the adult rat hypothalamus (mostly the core hypothalamus) in two sagittal planes.

11 Summary

Four collections of thymidine radiograms were analyzed. *Short-survival radiograms* were used to study regional differences in cell proliferation in the ventral third ventricle (hypothalamic) neuroepithelium and, where possible, identify the sites of origin (neuroepithelial zones) of neurons of the subdivisions of the hypothalamus. *Sequential radiograms* were examined to trace the migratory trajectory of neurons and the pattern of their settling. *Fetal long-survival radiograms* provided information about the neurogenetic organization of the different hypothalamic areas and nuclei, and aided us in delineating their boundaries, including those of several hitherto undescribed nuclei and subnuclei. Finally, *adult long-survival radiograms* from an earlier study provided quantitative data about the chronology of neurogenesis throughout the hypothalamus.

All diencephalic structures situated beneath the thalamus and arising from the ventral portion of the third ventricle neuroepithelium are considered to belong to the hypothalamus. These include some structures (zona incerta, entopeduncular nucleus, and subthalamic nucleus) often considered to be part of the ventral thalamus or subthalamus. On the basis of embryonic landmarks along the third ventricle from rostral to caudal, the hypothalamus is divided into *five divisions:* the preoptic area, the anterior hypothalamus, the ventral region of the posterior hypothalamus, the mammillary body, and the caudal region of the posterior hypothalamus. In the lateromedial plane, the hypothalamus is divided into *three tiers:* the lateral hypothalamus, the core hypothalamus, and the midline hypothalamus. These three tiers arise from three successive (partially overlapping) waves of neurons of the hypothalamic neuroepithelium as it recedes from lateral to medial. The *subdivisions* (areas, nuclei, and subnuclei) of the hypothalamus constituting the three tiers are considered to represent three structurally and functionally separate components of the hypothalamus.

Development of the Reticular Hypothalamus. The reticular hypothalamus consists of the lateral hypothalamus, the entopeduncular nucleus, and the zona incerta. The neurons of these three structures constitute the first wave of differentiating cells that leave the hypothalamic third ventricle while it still occupies a far-lateral position. The neurons of the *lateral hypothalamus* are generated between days E13 and E15. They originate in a segment of the neuroepithelium situated caudal and beneath the presumptive neuroepithelium of the zona incerta. The fibers of the medial forebrain bundle begin to penetrate the area after day E15. The bulk of the neurons of the *entopeduncular nucleus* are produced before the morning of day E14. Evidence is presented that, unlike the later-generated neurons of the adjacent external globus pallidus, the entopeduncular neurons derive from the third ventricle neuroepithelium from a site dorsal to that of the zona incerta. Entopeduncular neurons begin to settle by day E14, before penetration of this region by fibers of the internal capsule. The neurons of the *zona incerta* are produced between days E13 and E15. The earliest-produced neurons settle medially and the latest-produced neurons laterally. The site of origin of zona incerta neurons seems to be located beneath the reticular protuberance that separates the thalamic and hypothalamic neuroepithelium.

Development of the Preoptic Area. The preoptic area is composed of structures of the core hypothalamus that derive from the portion of the third ventricle neuroepithelium that extends from the tip of the preoptic recess rostrally to the anterior wall of the optic fissure caudally. A rostral and a caudal portion of the preoptic area is distinguished and each is subdivided into many regions on the basis of neurogenetic criteria. The rostral portion consists of the *anterolateral, superior,* and *anteromedial preoptic* regions. The neurons of the anteromedial region are produced with a peak on day E13, the neurons of the superior region with a peak on day E14, and the neurons of the anteromedial region mostly between days E15 and E17. The caudal portion of the preoptic area consists of the *posterolateral, superior, posterodorsal, posteroventral* and *intermediate preoptic* regions. The neurons of the posterolateral region are produced first, followed by the neurons of the superior and intermediate regions, and ending with the neurons of the posterodorsal and posteroventral regions. The pattern of cell migration and settling of neurons is different in the rostral and caudal portions of the preoptic area. The anterolateral and posterolateral regions of the preoptic area may represent rostral extensions of the lateral hypothalamus.

Development of the Anterior Hypothalamus. The anterior hypothalamic structures derive from the portion of the third ventricle neuroepithelium extending from the posterior wall of the optic fissure rostrally to the infundibular recess caudally. The anterior hypothalamus is divided for convenience into a rostral and a caudal region. The main components of the rostral anterior hypothalamus are several anterior hypothalamic nuclei, and the paraventricular and supraoptic nuclei. The major components of the caudal anterior hypothalamus are the dorsomedial and ventromedial nuclei.

The three components of the rostrally situated anterior hypothalamic nuclei are the *anterobasal, anteroventral,* and *anterodorsal nuclei.* The neurons of the anterobasal nucleus are produced first, those of the anteroventral nucleus next, and those of the anterodorsal nucleus last. The site of origin of these neurons has been tentatively identified. The *paraventricular nucleus* is subdivided into three magnocellular components, the *anterolateral, medial,* and *posterior subnuclei,* and the parvocellular *periventricular subnucleus.* Peak production of neurons of the anterolateral subnucleus is on day E14, of the medial and posterior subnuclei on day E15, and of the periventricular subnucleus on day E16. The magnocellular neurons derive from a portion of the "diamond-shaped" neuroepithelium dorsally and the parvocellular neurons from a more ventrally situated neuroepithelium. The magnocellular neurons of the *supraoptic nucleus* are produced over a 3-day period between days E13 and E15. They appear to originate in the same neuroepithelial zone as the magnocellular neurons of the paraventricular nucleus. The settling of the supraoptic neurons begins laterally on day

E15 in a rostral to caudal sequence, before the arrival of optic tract fibers, and is completed thereafter on day E18. In the maturing brain the earliest-produced supraoptic neurons are located dorsolaterally in a retrochiasmatic position, and the latest-generated neurons ventromedially in a juxtachiasmatic position. The possibility that the early-generated magnocellular neurons synthesize vasopressin and the late-generated magnocellular neurons synthesize oxytocin is discussed.

The *ventromedial nucleus* is divided on the basis of neurogenetic criteria into three parts: the *pars ventralis, pars dorsalis,* and *pars basalis.* The neurons of the ventromedial nucleus originate in a discrete neuroepithelial zone between days E13 and E17, and settle in an oblique ventrolateral-to-dorsomedial sequence, with the neurons of the pars basalis being composed of the last-produced neurons. The dorsomedial nucleus is subdivided into a *pars dorsalis, pars ventralis,* and *pars basalis* on the basis of a neurogenetic gradient that is the mirror image of that seen in the ventromedial nucleus. Neuron production in this nucleus is between days E14 and E17 and apparently the cells derive from a neuroepithelial zone that is distinct from that of the ventromedial nucleus.

Development of the Ventral Region of the Posterior Hypothalamus. The components of the ventral posterior hypothalamus derive from the portion of the third ventricle neuroepithelium that extends from the infundibular recess anteriorly to the mammillary recess posteriorly. The three major structures of this division are the subthalamic nucleus, the tuberal magnocellular nucleus and the premammillary nuclei. The *subthalamic nucleus* (of Luys), which most anatomists do not consider to be part of the hypothalamus, derives from a neuroepithelial zone in a anterodorsal region of the mammillary recess, rostral to the site of origin of the tuberal magnocellular neurons. The bulk of its neurons are produced on days E14 and E15. Its migrating cells follow a semicircular route dorsolaterally and settle in an outside-in pattern. The earliest-produced subthalamic neurons become situated dorsolaterally and its last-produced elements ventromedially. On the basis of neurogenetic criteria an *accessory subthalamic nucleus* has been identified. The neurons of the *tuberal magnocellular nucleus* are produced between days E15 and E18, with peaks on days E16 and E17. They originate from the caudal tip of the mammillary recess, dorsal to the site of production of the arcuate neurons. The neurons of the *ventral premammillary nucleus* are generated predominantly on day E14, with a smaller proportion added on day E15, and the neurons of the *dorsal premammillary nucleus* are produced mostly on days E15 and E16, with a small proportion added on day E17. The site of origin of neurons of the premammillary nuclei appears to be a neuroepithelial zone dorsal to the source of arcuate neurons.

Development of the Dorsal Region of the Posterior Hypothalamus. The components of the dorsal posterior hypothalamus derive from the portion of the third ventricle neuroepithelium situated above the mammillary recess and extending to the inner diencephalic flexure. Its major components are the supramammillary nucleus and the posterior area nuclei. The *supramammillary nucleus* is divisible into a *pars medialis* and a *pars lateralis*. The pars medialis neurons are generated predominantly on days E15 and E16, the pars lateralis neurons on days E16 and E17. This gradient (medial-to-lateral) is the reverse of that found in the components of the mammillary body. Three components of the *posterior area nuclei* are distinguished: the *dorsocaudal nucleus*, the *perifornical nucleus*, and the *suprafornical nucleus*. The neurons of the perifornical nucleus are generated first, the dorsocaudal nucleus next, and the suprafornical nuclei last. It is assumed that these three nuclei stem from a neuroepithelial zone dorsal to the site of origin of the premammillary nuclei.

Development of the Mammillary Body. The neurons of the mammillary body originate from third ventricle neuroepithelium capping the mammillary recess. On the basis of cytological and neurogenetic criteria, the mammillary body is divided into a *lateral nucleus* and a *medial nucleus*. The medial nucleus has two components, the *central* and *intermediate subnuclei*. The intermediate subnucleus is further divided into a *pars dorsalis, pars ventralis,* and a *diffuse region*. The neurons of these divisions of the mammillary body are produced in a sequence and within each a neurogenetic gradient has been observed. The lateral mammillary neurons are produced first, predominantly on days E13 and E15, and settle in a dorsolateral-to-ventromedial sequence. The pars dorsalis of the intermediate subnucleus is comprised of the next generation of neurons and the gradient here is lateral-to-medial. Third in sequence is the central subnucleus, and in it the gradient is likewise lateral-to-medial, with a few of the medial cells produced on day E17. The neurons of the pars ventralis of the intermediate subnucleus are fourth in order of production and they settle in a dorsolateral-to-ventromedial sequence. The cells that are still forming on days E18 and E19 settle in the diffuse region of the medial mammillary nucleus.

Development of the Midline Hypothalamus. The neurons of the midline hypothalamus represent the last wave of hypothalamic neurons that settle in the vicinity of the medially receding third ventricle neuroepithelium. Its small neurons have been implicated in the synthesis and transport of pituitary-releasing hormones. The midline hypothalamus has two condensed components, the suprachiasmatic nucleus and the arcuate nucleus, and a diffuse component, the periventricular field. The *suprachiasmatic* nucleus is divided into three neurogenetic parts: the *ventrolateral, dorsomedial,* and *basal subnuclei*. Neurogenesis ends in the ventrolateral subnucleus by day E16 and in the dorsomedial and basal subnuclei on day E17. The suprachiasmatic neuroepithelial zone forms a shallow sulcus caudal to the optic recess. The neurons of the *arcuate nucleus* are produced over an extended period with a distinct rostral-to-caudal gradient. Rostrally neuron production ends on day E17 but caudally some arcuate neurons are produced as late as day E19. A less-pronounced lateral-to-medial neurogenetic gradient is attributable to the progressive withdrawal of the putative arcuate neuroepithelium toward the midline. The small neurons of the *periventricular field* are scattered near the ependymal lining of the third ventricle from the preoptic rostrally to the posterior hypothalamus caudally. In some regions the neurons of the periventricular field form a continuum with medial portions of named hypothalamic nuclei, but may be distinguished from the latter by the later production of their neurons. A chronological link was noted between the regional production of periventricular neurons and the cells of the specialized lining of the hypothalamic third ventricle.

Development of the Nonneuronal Elements of the Hypothalamus. Three specialized linings of the hypothalamic third ventricle are identified: the *convoluted ependyma,* the *laminated epithelium,* and the *tanycytic epithelium.* The cells of these linings are produced after the cessation of neurogenesis through the perinatal and early postnatal periods. The convoluted ependyma may be derived from the transformed neuroepithelium in the region of the paraventricular nucleus; the laminated epithelium from that of the dorsomedial nucleus; and the two components of the tanycytic epithelium may be related to the germinal sites of the ventromedial and arcuate nuclei. Evidence is presented that the glial cells of the hypothalamus derive from dispersed, locally multiplying cells. This regional proliferative activity increases after the cessation of neurogenesis.

Theoretical Remarks. The examination of hypothalamic neurogenesis suggests that there is a mosaic of discrete cell lines in the ventral third ventricle neuroepithelium, each characterized by a special chronology of precursor cell production and spatial distribution. Much further work is needed to translate this approach into an accurate map of the neuroepithelial zones of the germinal matrix of the hypothalamus. Future research may also establish to what extent the precursor cells of a discrete neuroepithelial zone express the biochemical properties of the differentiating neurons they are destined to produce.

References

Abe K, Kroning J, Greer MA, Critchlow V (1979) Effects of destruction of the suprachiasmatic nuclei on the circadian rhythms in plasma corticosterone, body temperature, feeding, and plasma thyrotropin. Neuroendocrinology 29:119–131

Afifi A, Kaelber WW (1965) Efferent connections of the substantia nigra in the cat. Exp Neurol 11:474–482

Afsharpour S (1985) Topographical projections of the cerebral cortex to the subthalamic nucleus. J Comp Neurol 236:14–28

Aghajanian GK, Bloom FE, Sheard MH (1969) Electron microscopy of degeneration within the serotonin pathway of rat brain. Brain Res 13:266–273

Altman J (1966) Proliferation and migration of undifferentiated precursor cells in the rat during postnatal gliogenesis. Exp Neurol 16:263–278

Altman J, Bayer SA (1978a) Development of the diencephalon in the rat. I. Autoradiographic study of the time of origin and settling patterns of neurons of the hypothalamus. J Comp Neurol 182:945–972

Altman J, Bayer SA (1978b) Development of the diencephalon in the rat. II. Correlation of the embryonic development of the hypothalamus with the time of origin of its neurons. J Comp Neurol 182:973–994

Altman J, Bayer SA (1978c) Development of the diencephalon in the rat. III. Ontogeny of the specialized ventricular linings of the hypothalamic third ventricle. J Comp Neurol 182:995–1016

Altman J, Bayer SA (1979a) Development of the diencephalon in the rat. IV. Quantitative study of the time of origin of neurons and the internuclear chronological gradients in the thalamus. J Comp Neurol 188:455–472

Altman J, Bayer SA (1979b) Development of the diencephalon in the rat. V. Thymidine-radiographic observations on internuclear and intranuclear gradients in the thalamus. J Comp Neurol 188:473–500

Altman J, Bayer SA (1979c) Development of the diencephalon in the rat. VI. Re-evaluation of the embryonic development of the thalamus on the basis of thymidine-radiographic datings. J Comp Neurol 188:501–524

Altman J, Bayer SA (1981) Development of the brain stem in the rat. V. Thymidine-radiographic study of the time of origin of neurons in the midbrain tegmentum. J Comp Neurol 198:677–716

Altman J, Bayer SA (1984) The development of the rat spinal cord. Springer, Berlin Heidelberg New York Tokyo (Advances in anatomy, embryology and cell biology, vol 85)

Anand BK, Brobeck JR (1951) Hypothalamic control of food intake in rats and cats. Yale J Biol Med 24:123–140

Anderson CH (1978) Time of neuron origin in the anterior hypothalamus of the rat. Brain Res 154:119–122

Armstrong WE, Warach S, Hatton GI, McNeill TH (1980) Subnuclei in the rat hypothalamic paraventricular nucleus: a cytoarchitectural, horseradish peroxidase and immunocytochemical analysis. Neuroscience 5:1931–1958

Auer J (1951) Postnatal cell differentiation in the hypothalamus of the hamster. J Comp Neurol 95:17–41

Baker BL, Dermody WC, Reel JR (1975) Distribution of gonadotropin-releasing hormone in the rat brain as observed with immunocytochemistry. Endocrinology 97:125–135

Barfield RJ, Chen JJ (1977) Activation of estrous behavior in ovariectomized rats by intracerebral implants of estradiol benzoate. Endocrinology 101:1716–1725

Bargmann W (1949) Über die neurosekretorische Verknüpfung von Hypothalamus und Neurohypophyse. Z Zellforsch 34:610–634
Barraclough CA (1973) Sex steroid regulation of reproductive neuroendocrine processes. In: Creep RO (ed) Handbook of physiology, endocrinology II, part I. Waverley, Baltimore, pp 29–56
Barry J, Dubois MP, Carrette B (1974) Immunofluorescence study of the preoptic-infundibular LRF neurosecretory pathway in the normal, castrated or testosterone-treated male guinea pig. Endocrinology 95:1416–1423
Bayer SA (1979a) The development of the septal region in the rat. I. Neurogenesis examined with ^{3}H-thymidine autoradiography. J Comp Neurol 183:89–106
Bayer SA (1979b) The development of the septal region in the rat. II. Morphogenesis in normal and x-irradiated embryos. J Comp Neurol 183:107–120
Bayer SA (1985) Neurogenesis of the magnocellular basal telencephalic nuclei in the rat. Int J Devel Neurosci 3:229–243
Beaudet A, Descarries L (1979) Radioautographic characterization of a serotonin-accumulating nerve cell group in adult rat hypothalamus. Brain Res 160:231–243
Bennett-Clarke C, Romagnano MA, Joseph SA (1980) Distribution of somatostatin in the rat brain: telencephalon and diencephalon. Brain Res 188:473–486
Bergquist HJ, Källén AJB (1954) Notes on the early histogenesis and morphogenesis of the central nervous system in vertebrates. J Comp Neurol 100:627–659
Björklund A, Moore RY, Nobin A, Stenevi U (1973) The organization of tuberohypophyseal and reticulo-infundibular catecholamine neuron systems in the rat brain. Brain Res 51:171–191
Bleier R (1971) The relations of ependyma to neurons and capillaries in the hypothalamus: a Golgi-Cox study. J Comp Neurol 142:439–464
Bleier R, Byne W, Siggelkow I (1982) Cytoarchitectonic sexual dimorphisms of the medial preoptic and anterior hypothalamic areas in guinea pig, rat, hamster, and mouse. J Comp Neurol 212:118–130
Bleier R, Cohn P, Siggelkow IR (1979) A cytoarchitectonic atlas of the hypothalamus and hypothalamic third ventricle of the rat. In: Morgane PJ, Panksepp J (eds) Anatomy of the hypothalamus. Dekker, New York, pp 137–220 (Handbook of the hypothalamus, vol 1)
Bobillier P, Seguin S, Degueurce A, Lewis BD, Pujol JF (1979) The efferent connections of the nucleus raphe centralis superior in the rat as revealed by radioautography. Brain Res 166:1–8
Brawer JR (1971) The role of the arcuate nucleus in the brain-pituitary-gonad axis. J Comp Neurol 143:411–446
Brobeck JR, Tepperman J, Long CNH (1943) Experimental hypothalamic hyperphagia in the albino rat. Yale J Biol Med 15:831–853
Brownstein MJ, Palkovits J, Saavedra JM, Bassiri RM, Utiger RD (1974) Thyrotropin-releasing hormone in specific nuclei of rat brain. Science 185:267–269
Brügger M (1943) Fresstrieb als hypothalamisches Symptom. Helv Physiol Pharmacol Acta 1:183–198
Bruni JE, Montemurro DG, Clattenburg RE, Singh RP (1972) A scanning electron microscopic study of the ependymal surface of the third ventricle of the rabbit, rat, mouse and human brain. Anat Rec 174:407–420
Bugnon C, Bloch B, Lenys D, Gouget A, Fellmann D (1979) Comparative study of the neuronal populations containing beta-endorphin, corticotropin and dopamine in the arcuate nucleus of the rat hypothalamus. Neurosci Lett 14:43–48
Buijs RM (1978) Intra- and extra-hypothalamic vasopressin and oxytocin pathways in the rat. Cell Tissue Res 192:423–435
Carpenter MB, Batton RB, Carleton SC, Keller JT (1981) Interconnections and organization of pallidal and subthalamic nucleus neurons in the monkey. J Comp Neurol 197:579–603
Carpenter MB, Whittier JR, Mettler FA (1950) Analysis of choreoid hyperkinesia in the rhesus monkey. J Comp Neurol 92:293–332
Carter DA, Fibiger HC (1977) Ascending projections of presumed dopamine-containing neurons in the ventral tegmentum of the rat as demonstrated by horseradish peroxidase. Neuroscience 2:569–576
Carter DA, Fibiger HC (1978) The projections of the entopeduncular nucleus and globus

pallidus in rat as demonstrated by autoradiography and horseradish peroxidase histochemistry. J Comp Neurol 177:113–124
Choy VJ, Watkins WB (1979) Maturation of the hypothalamo-neurohypophysial system. I. Localization of neurophysin, oxytocin and vasopressin in the hypothalamus and neural lobe of the developing rat brain. Cell Tissue Res 197:325–336
Christ JF (1969) Derivation and boundaries of the hypothalamus, with atlas of hypothalamic grisea. In: Haymaker et al. (eds) The hypothalamus. Thomas, Springfield, pp 13–60
Christensen LW, Clemens LG (1974) Intrahypothalamic implants of testosterone or estradiol and resumption of masculine sexual behavior in long-term castrated male rats. Endocrinology 95:984–990
Clark Le Gros WE, Beattie J, Riddoch G, Dott NM (1938) In: Clark Le Gros WE et al. (eds) The hypothalamus: morphological, functional, clinical and surgical aspects. Oliver and Boyd, Edinburgh, 213 pp
Clemens JA, Smalstig EB, Sawyer BD (1976) Studies on the role of the preoptic area in the control of reproductive function in the rat. Endocrinology 99:728–735
Coggeshall RE (1964) A study of diencephalic development in the albino rat. J Comp Neurol 122:241–269
Colmant HJ (1967) Über die Wandstruktur des dritten Ventrikels der Albinoratte. Histochemie 11:40–61
Conrad LCA, Pfaff DW (1976a) Efferents from medial basal forebrain and hypothalamus in the rat. I. An autoradiographic study of the medial preoptic area. J Comp Neurol 169:185–220
Conrad LCA, Pfaff DW (1976b) Efferents from medial basal forebrain and hypothalamus in the rat. II. An autoradiographic study of the anterior hypothalamus. J Comp Neurol 169:221–262
Cowan WM, Guillery RW, Powell TPS (1964) The origin of the mammillary peduncle and other hypothalamic connexions from the midbrain. J Anat (Lond) 98:345–363
Cruce JAF (1977) An autoradiographic study of the descending connections of the mammillary nuclei of the rat. J Comp Neurol 176:631–644
Defendini R, Zimmerman EA (1978) The magnocellular neurosecretory system of the mammalian hypothalamus. In: Reichlin S, Baldessarini RJ (eds) The hypothalamus. Raven, New York, pp 137–152
de Groot J (1959) The rat hypothalamus in stereotaxic coordinates. J Comp Neurol 113:389–400
Deniau JM, Hammond C, Chevalier G, Feger J (1978) Evidence for branched subthalamic nucleus projections to substantia nigra, entopeduncular nucleus and globus pallidus. Neurosci Lett 9:117–121
De Olmos JS, Ingram WR (1972) The projection field of the stria terminalis in the rat brain. An experimental study. J Comp Neurol 146:303–334
De Vito JL, Smith OA (1964) Subcortical projections of the prefrontal lobe of the monkey. J Comp Neurol 123:413–424
de Vries GJ, Buijs RM, van Leeuwen FW, Caffé AR, Swaab DF (1985) The vasopressinergic innervation of the brain in normal and castrated rats. J Comp Neurol 233:236–254
Diepen R (1962) Der Hypothalamus. Springer, Berlin Heidelberg New York (Handbuch der Mikroskopischen Anatomie des Menschen, vol 4, part 17)
Dierickx K (1980) Immunocytochemical localization of the vertebrate cyclic nonapeptide neurohypophyseal hormones and neurophysins. Int Rev Cytol 62:119–185
Dierickx K, Vandesande F (1979) Immunocytochemical localization of somatostatin-containing neurons in the rat hypothalamus. Cell Tissue Res 201:349–359
Edinger L (1896) Vorlesungen über den Bau der Nervösen Zentralorgane des Menschen. Vogel, Leipzig
Emery DE, Sachs BD (1976) Copulatory behavior in male rats with lesions in the bed nucleus of the stria terminalis. Physiol Behav 17:803–806
Eskay RL, Giradu P, Oliver C, Brownstein MJ (1979) Distribution of alpha-melanocyte-stimulating hormone in the rat brain. Brain Res 178:55–67
Everett JW (1964) Central neural control of reproductive functions of the adenohypophysis. Physiol Rev 44:373–431
Filion M, Harnois C (1978) A comparison of projections of entopeduncular neurons to the thalamus, the midbrain and the habenula in the cat. J Comp Neurol 181:763–780

Forel A (1877) Untersuchungen über die Haubenregion und ihre oberen Verknüpfungen im Gehirn des Menschen und der Säugethiere, mit Beiträgen zu den Methoden der Gehirnforschung. Arch Psychiatr 7:393–495
Fox CA, Hillman EE, Siegesmund KA, Sether LA (1966) The primate globus pallidus and its feline and avian homologues: a Golgi and electron microscopic study. In: Hassler R, Stephan H (eds) Evolution of the forebrain. Thieme, Stuttgart, pp 237–248
Frohman LA, Bernardis LL (1971) Growth hormone and insulin levels in weanling rats with ventromedial hypothalamic lesions. Endocrinology 82:1125–1132
Frohman LA, Bernardis LL, Stachura M (1974) Factors modifying plasma insulin and glucose responses to ventromedial hypothalamic stimulation. Metabolism 23:1047–1056
Fry JF, Cowan WM (1972) A study of retrograde cell degeneration in the lateral mammillary nucleus of the cat with special reference to the role of axonal branching in the preservation of the cell. J Comp Neurol 144:1–24
Fuxe K (1965) Evidence for the existence of monoamine containing neurons in the central nervous system. Acta Physiol Scand 64 (Suppl 247):39–85
Gilbert MS (1935) The early development of the human diencephalon. J Comp Neurol 62:81–115
Gorski RA, Gordon JH, Shryne JE, Southam AM (1978) Evidence for a morphological sex difference within the medial preoptic area of the rat brain. Brain Res 148:333–346
Gray GD, Södersten P, Tallentire D, Davidson JM (1978) Effects of lesions in various structures of the suprachiasmatic-preoptic region on LH regulation and sexual behavior in female rats. Neuroendocrinology 25:174–191
Greenwood RS, Godar SE, Reaves TA, Hayward JN (1981) Cholecystokinin in hippocampal pathways. J Comp Neurol 203:335–350
Grizzle WE, Dallman MF, Schramm LP, Gann DS (1974) Inhibitory and facilitatory hypothalamic areas mediating ACTH release in the cat. Endocrinology 95:1450–1461
Grofová I (1969) Experimental demonstration of a topical arrangement of pallidosubthalamic fibers in the cat. Psychiatr Neurol Neurochir 72:53–59
Grossman SP (1976) Neuroanatomy of food and water intake. In: Novin D et al. (eds) Hunger: basic mechanisms and clinical implications. Raven, New York, pp 51–59
Grünthal E (1930) Vergleichend anatomische und entwicklungsgeschichtliche Untersuchungen über die Zentren des Hypothalamus der Säuger und des Menschen. Arch Psychiatr 90:216–267
Gurdjian ES (1927) The diencephalon of the albino rat. J Comp Neurol 43:1–114
Haglund L, Swanson LW, Köhler C (1984) The projection of the supramammillary nucleus to the hippocampal formation: an immunohistochemical and anterograde transport study with the lectin PHA-L in the rat. J Comp Neurol 229:171–185
Halász B (1969) The endocrine effect of isolation of the hypothalamus from the rest of the brain. In: Ganong WF, Martini L (eds) Frontiers in neuroendocrinology. Oxford University Press, New York, pp 307–342
Halász B, Kosaras B, Lengváry I (1972) Ontogenesis of the neurovascular link between the hypothalamus and the anterior pituitary in the rat. In: Knigge KM et al. (eds) Brain-endocrine interaction: median eminence, structure and function. Karger, Basel, pp 27–34
Handelmann GE, O'Donohue TL, Forrester D, Cook W (1983) Alpha-melanocyte stimulating hormone facilitates learning of visual but not auditory discriminations. Peptides 4:145–148
Harris MC, Sanghera M (1974) Projection of medial basal hypothalamic neurons to the preoptic anterior hypothalamic areas and the paraventricular nucleus in the rat. Brain Res 81:401–411
Hartmann-von Monakow K, Akert K, Künzle H (1978) Projections of the precentral motor cortex and other cortical areas of the frontal lobe to the subthalamic nucleus in the monkey. Exp Brain Res 33:395–403
Hatton GI, Hutton UE, Hoblitzell ER, Armstrong WE (1976) Morphological evidence for two populations of magnocellular elements in the rat paraventricular nucleus. Brain Res 108:187–193
Heimer L, Larsson K (1966/1967) Impairment of mating behavior in male rats following lesions in the preoptic-anterior hypothalamic continuum. Brain Res 3:248–263
Hendrickson AE, Wagoner N, Cowan WM (1972) An autoradiographic and electron microscopic study of retino-hypothalamic connections. Z Zellforsch 135:1–26

Herrick CJ (1910) The morphology of the forebrain in amphibia and reptilia. J Comp Neurol 20:413–547
Hetherington AW, Ranson SW (1942) The spontaneous activity and food intake of rats with hypothalamic lesions. Am J Physiol 136:609–617
His W (1893) Vorschläge zur Entheilung des Gehirns. Arch Anat Entw Gesch 172–179
Hökfelt T, Fuxe K, Goldstein M, Johansson O (1974) Immunohistochemical evidence for the existence of adrenaline neurons in the rat brain. Brain Res 66:235–252
Holstege G, Meiners L, Kan T (1985) Projections of the bed nucleus of the stria terminalis to the mesencephalon, pons and medulla oblongata in the cat. Exp Brain Res 58:379–391
Hosoya Y, Matsushita M (1979) Identification and distribution of the spinal and hypophyseal projection neurons in the paraventricular nucleus of the rat. A light and electron microscopic study with the horseradish peroxidase method. Exp Brain Res 35:315–331
Hyyppä M (1969) Differentiation of the hypothalamic nuclei during ontogenetic development in the rat. Z Anat Entw Gesch 129:41–52
Ibuka N, Kawamura H (1975) Loss of circadian rhythm in sleep-wakefulness cycle in the rat by suprachiasmatic nucleus lesions. Brain Res 96:76–81
Ifft JD (1972) An autoradiographic study of the time of final division of neurons in rat hypothalamic nuclei. J Comp Neurol 144:193–204
Jacobowitz DM (1975) Fluorescence microscopic mapping of CNS norepinephrine systems in the rat forebrain. In: Stumpf WE, Grant LD (eds) Anatomical neuroendocrinology. Karger, Basel, pp 368–380
Jacobowitz DM, Schulte H, Chrousos GP, Loriaux DL (1983) Localization of GRF-like immunoreactive neurons in the rat brain. Peptides 4:521–524
Jacobson CD, Shryne JE, Shapiro F, Gorski RA (1980) Ontogeny of the sexually dimorphic nucleus of the preoptic area. J Comp Neurol 193:541–548
Joseph SA (1980) Immunoreactive adrenocorticotropin in rat brain: a neuroanatomical study using antiserum generated against synthetic ACTHl-39. Am J Anat 158:533–548
Joseph SA, Knigge KM (1978) The endocrine hypothalamus: recent anatomical studies. In: Reichlin S, Baldessarini RJ (eds) The hypothalamus. Raven, New York, pp 15–47
Kaelber WW, Smith TB (1979) Projections of the zona incerta in the cat, with stimulation controls. Exp Neurol 63:177–200
Kahle W (1956) Zur Entwicklung des menschlichen Zwischenhirnes. Studien über die Matrixphasen und die örtlichen Reifungsunterschiede im embryonalen menschlichen Gehirn. II. Mitteilung. Dtsch Z Nervenheilk 175:259–318
Kahn D, Abrams GM, Zimmerman EA, Carraway R, Leeman S (1980) Neurotensin neurons in the rat hypothalamus: an immunocytochemical study. Endocrinology 107:47–54
Kelly MJ, Ronnekleiv OK, Eskay RL (1982) Immunocytochemical localization of luteinizing hormone-releasing hormone in neurons in the medial basal hypothalamus of the female rat. Exp Brain Res 48:97–106
Kennedy GC (1963) Hypothalamic control of energy balance and reproductive cycle in rat. J Physiol (Lond) 166:395–407
Keyser A (1972) The development of the diencephalon of the Chinese hamster. Acta Anat 83 [Suppl 59]:1–178
Khachaturian H, Sladek JR (1980) Simultaneous monoamine histofluorescence and neuropeptide immunocytochemistry. III. Ontogeny of catecholamine varicosities and neurophysin neurons in the rat supraoptic and paraventricular nuclei. Peptides 1:77–95
Kim R, Nakano K, Jayaraman A, Carpenter MB (1976) Projections of the globus pallidus and adjacent structures: an autoradiographic study in the monkey. J Comp Neurol 169:263–290
King JC, Williams TH, Gerall AA (1974) Transformations of hypothalamic arcuate neurons. I. Changes associated with stages of the estrous cycle. Cell Tissue Res 153:497–515
King MB, Hoebel B (1968) Killing by brain stimulation in the rat. Communic Behav Biol 2:173–177
Kita H, Chang HT, Kitai ST (1983a) The morphology of intracellularly labeled rat subthalamic neurons: a light microscopic analysis. J Comp Neurol 215:245–257
Kita H, Chang HT, Kitai ST (1983b) Pallidal inputs to subthalamus: intracellular analysis. Brain Res 264:255–265
Kita H, Oomura Y (1982) An HRP study of the afferent connections to rat medial hypothalamic region. Brain Res Bull 8:53–62

Kitai ST, Deniau JM (1981) Cortical inputs to the subthalamus: intracellular analysis. Brain Res 214:411–415

Knigge KM, Joseph SA, Nocton J (1981) Topography of the ACTH-immunoreactive neurons in the basal hypothalamus of the rat brain. Brain Res 216:333–341

Kobayashi T, Kobayashi T, Yamamoto K, Kaibara M, Ajika K (1968) Electron microscopic observation on the hypothalamo-hypophyseal system in rats. IV. Ultra-fine structure of the developing median eminence. Endocrinol Japan 15:337–363

Kobayashi H, Wada M, Uemura H, Ueck M (1972) Uptake of peroxidase from the third ventricle by ependymal cells of the median eminence. Z Zellforsch 127:545–551

Köhler C, Haglund L, Swanson LW (1984) A diffuse alpha MSF-immunoreactive projection to the hippocampus and spinal cord from individual neurons in the lateral hypothalamic area and zona incerta. J Comp Neurol 223:501–514

König JFR, Klippel RA (1963) The rat brain. Williams and Wilkins, Baltimore

Koritsánszky S (1979) Cyto- and synaptogenesis in the arcuate nucleus of the rat hypothalamus during fetal and early postnatal life. Cell Tissue Res 200:135–146

Kozlowski GP, Scott DE, Dully GK (1973) Scanning electron microscopy of the third ventricle of sheep. Z Zellforsch 136:169–176

Krettek JE, Price JL (1978) Amygdaloid projections to subcortical structures within the basal forebrain and brainstem in the rat and cat. J Comp Neurol 178:225–254

Krieg WJS (1932) The hypothalamus of the albino rat. J Comp Neurol 55:19–89

Künzle H (1978) An autoradiographic analysis of the efferent connection from premotor and adjacent prefrontal regions (areas 6 and 9) in the *Macaca fascicularis*. Brain Behav Evol 15:185–234

Kuhlenbeck H (1923) Weitere Mitteilungen über die Genese der Basalganglion. Über die sog. Ganglienhügel. Anat Anz 58:49–74

Kuhlenbeck H (1973) The central nervous system of vertebrates, vol 3, part II. Karger, Basel

Kuhlenbeck H (1977) The central nervous system of vertebrates, vol 5, part I. Karger, Basel

Larsen KD, McBride RL (1979) The organization of feline entopeduncular nucleus projections: anatomical studies. J Comp Neurol 184:293–308

Law T, Meagher W (1958) Hypothalamic lesions and sexual behavior in the female rat. Science 128:1626–1627

Lechan RM, Nestler JL, Jacobson S (1982) The tuberoinfundibular system of the rat as demonstrated by immunohistochemical localization of retrogradely transported wheat germ agglutinin (WGA) from the median eminence. Brain Res 245:1–15

Leonhardt H (1968) Bukettförmige Strukturen im Ependym de regio hypothalamica des II. Ventrikels beim Kaninchen. Zur Neurosekretions- und Rezeptorenfrage. Z Zellforsch 88:297–317

Lisk RD (1962) Diencephalic placement of estradiol and sexual receptivity in the female rat. Am J Physiol 203:493–496

Lisk RD (1967) Neural localization for androgen activation of copulatory behavior in the male rat. Endocrinology 80:754–761

Lisk RD (1968) Copulatory activity of the male rat following placement of preoptic-anterior hypothalamic lesions. Exp Brain Res 5:306–313

Luyendijk W (1944) Studies on the hypothalamic region of the diencephalon. Acta Neerl Morph 5:248–261

Magoun HW (1950) Caudal and cephalic influences of the brain stem reticular formation. Physiol Rev 30:459–474

Mai JK, Junger E (1977) Quantitative autoradiographic light- and electron microscopic studies on the retinohypothalamic connections in the rat. Cell Tissue Res 183:221–237

Makara GB, Palkovits M, Antoni FA, Kiss JZ (1983) Topography of the somatostatin-immunoreactive fibers to the stalk-median eminence of the rat. Neuroendocrinology 37:1–8

Malsbury CW (1971) Facilitation of male rat copulatory behavior by electrical stimulation of the preoptic area. Physiol Behav 7:797–805

Malone EF (1910) Über die Kerne des menschlichen Diencephalons. Abhandl Königl Preuss Akad Wissensch, Berlin

Marshall JF, Teitelbaum P (1974) Further analysis of sensory inattention following lateral hypothalamic damage in rats. J Comp Physiol Psychol 86:375–395

Martin JP (1927) Hemichorea resulting from a local lesion of the brain (syndrome of body of Luys). Brain 50:637–651

Matthews D, Edwards DA (1977) The ventromedial nucleus of the hypothalamus and hormonal arousal of sexual behaviors in the female rat. Horm Behav 8:40–51
Meibach RC, Siegel A (1975) The origin of fornix fibers which project to the mammillary bodies in the rat: a horseradish peroxidase study. Brain Res 88:508–512
Meibach RC, Siegel A (1977) Efferent connections of the hippocampal formation in the rat. Brain Res 124:197–224
Merchenthaler I, Kovács G, Lovász G, Sétáló G (1980) The preoptico-infundibular LH-RH tract of the rat. Brain Res 198:63–74
Merchenthaler I, Thomas CR, Arimura A (1984) Immunocytochemical localization of growth hormone releasing factor (GHRF)-containing structures in the rat brain using anti-rat GHRF-serum. Peptides 5:1071–1075
Meynert T (1872) Vom Gehirn der Säugethiere. In: Stricker (ed) Handbuch der Gewebelehre. Engelmann, Leipzig, pp:694–805
Millhouse OE (1969) A Golgi study of the descending medial forebrain bundle. Brain Res 15:341–363
Millhouse OE (1971) A Golgi study of the third ventricle tanycytes in the adult rodent brain. Z Zellforsch 121:1–13
Millhouse OE (1973) The organization of the ventromedial hypothalamic nucleus. Brain Res 55:71–87
Millhouse OE (1979) A Golgi anatomy of the rodent hypothalamus. In: Morgane PJ, Panksepp J (eds) Anatomy of the hypothalamus, vol 1. Dekker, New York, pp 221–265
Moore RY, Eichler VB (1972) Loss of a circadian adrenal corticosterone rhythm following suprachiasmatic lesions in the rat. Brain Res 42:201–206
Moore RY, Lenn NJ (1972) A retinohypothalamic projection in the rat. J Comp Neurol 146:1–14
Morin F (1950) An experimental study of hypothalamic connections in the guinea pig. J Comp Neurol 92:193–213
Naik DV (1975) Immunoreactive LH-RH neurons in the hypothalamus identified by light and fluorescent microscopy. Cell Tissue Res 157:423–436
Nauta HJW (1979) Projections of the pallidal complex: an autoradiographic study in the cat. Neuroscience 4:1853–1873
Nauta HJW, Cole M (1978) Efferent projections of the subthalamic nucleus: an autoradiographic study in monkey and cat. J Comp Neurol 180:1–16
Nauta WJH, Kuypers HGMJ (1958) Some ascending pathways in the brainstem reticular formation. In: Jasper HH et al. (eds) Reticular formation of the brain. Little, Brown, Boston, pp 3–30
Nauta WJH, Mehler WR (1966) Projections of the lentiform nucleus in the monkey. Brain Res 1:3–42
Nieuwenhuys R, Geeraedts LMG, Veening JG (1982) The medial forebrain bundle of the rat. I. General introduction. J Comp Neurol 206:49–81
Niimi K, Harada I, Kusaka Y, Kishi S (1962) The ontogenetic development of the diencephalon of the mouse. Tokushima J Exp Med 8:203–238
Nishizuka M (1979) Topography of the dopamine neurons in the arcuate nucleus of the mouse hypothalamus. Acta Anat 103:34–38
Numan M (1974) Medial preoptic area and maternal behavior in the female rat. J Comp Physiol Psychol 87:746–759
O'Donohue TL, Miller RL, Jacobowitz DM (1979) Identification, characterization and stereotaxic mapping of intraneuronal alpha-melanocyte stimulating hormone-like immunoreactive peptides in discrete regions of the rat brain. Brain Res 176:101–123
Palkovits M, Arimura A, Brownstein M, Schally AV, Saavedra JM (1974) Luteinizing hormone-releasing hormone (LH-RH) content of the hypothalamic nuclei in rat. Endocrinology 95:554–558
Palkovits M, Záborszky L (1979) Neural connections of the hypothalamus. In: Morgane PJ, Panksepp J (eds) Anatomy of the hypothalamus, vol 1. Dekker, New York, pp 379–509
Pasquier DA, Reinoso-Suarez F (1976) Direct projections from hypothalamus to hippocampus in the rat demonstrated by retrograde transport of horseradish peroxidase. Brain Res 108:165–169
Perkins MN, Stone TW (1980) Subthalamic projections to the globus pallidus: an electrophysiological study in the rat. Exp Neurol 68:500–511

Peterson RP (1966) Magnocellular neurosecretory centers in the rat hypothalamus. J Comp Neurol 128:181–189
Petras JM (1967) Some efferent connections of the motor and somatosensory cortex of simian primates and felid, canid and procyonid carnivores. Ann NY Acad Sci 167:469–505
Pfaff DW, Keiner M (1973) Atlas of estradiol-concentrating cells in the central nervous system of the female rat. J Comp Neurol 151:121–158
Pfaff DW, Sakuma Y (1979) Facilitation of the lordosis reflex of female rats from the ventromedial nucleus of the hypothalamus. J Physiol (Lond) 288:189–202
Piekut DT, Knigge KM (1984) Relationship of alpha MSH-specific neurons to the arcuate opiocortin neuronal system as determined by dual antigen immunocytochemical procedures. Peptides 5:1089–1095
Powers B, Valenstein ES (1972) Sexual receptivity: facilitation by medial preoptic lesions in femal rats. Science 175:1003–1005
Raisman G (1972) A second look at the parvicellular neurosecretory system. In: Knigge KM et al. (eds) Brain-endocrine interaction. Karger, Basel, pp 109–118
Raisman G, Field PM (1973) Sexual dimorphism in the neuropil of the preoptic area of the rat and its dependence on neonatal androgen. Brain Res 54:1–29
Rhodes CH, Morrell JI, Pfaff DW (1981) Immunohistochemical analysis of mangocellular elements in rat hypothalamus. J Comp Neurol 198:45–64
Ribak CE, Peters A (1975) An autoradiographic study of the projections from the lateral geniculate body of the rat. Brain Res 92:341–368
Ricardo JA (1981) Efferent connections of the subthalamic region in the rat. II. The zona incerta. Brain Res 214:43–60
Richter E (1965) Die Entwicklung des Globus Pallidus und des Corpus Subthalamicum. Springer, Berlin Heidelberg New York
Ridley PT, Brooks FP (1965) Gastric secretion following hypothalamic lesions producing hyperphagia. Am J Physiol 209:319–323
Rinvik E, Grofová I, Hammond C, Feger J, Deniau JM (1979) A study of the afferent connections to the subthalamic nucleus in the monkey and the cat using the horseradish peroxidase technique. Adv Neurol 24:53–69
Rioch DM, Wislocki GB, O' Leary JL (1940) Precis of preoptic, hypothalamic and hypophysial terminology with atlas. Assoc Res Nerv Ment Dis Proc 20:3–30
Rose J (1939) The cell structure of the mammillary body in mammals and in man. J Anat (Lond) 74:91–115
Rose JE (1942) The ontogenetic development of the rabbit's diencephalon. J Comp Neurol 77:61–129
Ruggiero DA, Baker H, Joh TH, Reis DJ (1984) Distribution of catecholamine neurons in the hypothalamus and preoptic region of mouse. J Comp Neurol 223:556–582
Rusak B, Zucker I (1979) Neural regulation of circadian rhythms. Physiol Rev 59:449–526
Sandman CA, Miller LH, Kastin AJ, Schally AV (1972) Neuroendocrine influence on attention and memory. J Comp Physiol Psychol 80:54–58
Saper CB, Swanson LW, Cowan WM (1976) The efferent connections of the ventromedial nucleus of the hypothalamus of the rat. J Comp Neurol 169:409–442
Saper CB, Swanson LW, Cowan WM (1979) An autoradiographic study of the efferent connections of the lateral hypothalamic area in the rat. J Comp Neurol 183:689–706
Sauer FC (1935) Mitosis in the neural tube. J Comp Neurol 62:377–405
Sawchenko PE, Swanson LW (1983) The organization of forebrain afferents to the paraventricular and supraoptic nuclei of the rat. J Comp Neurol 218:121–144
Scott DE, Kozlowski GP, Sheridan MN (1974) Scanning electron microscopy in the ultrastructural analysis of the mammalian cerebral ventricular system. Intern Rev Cytol 37:349–388
Segal M (1979) A potent inhibitory monosynaptic hypothalamo-hippocampal connection. Brain Res 162:137–141
Segal M, Landis S (1974) Afferents to the hippocampus of the rat studied with the method of retrograde transport of horseradish peroxidase. Brain Res 78:1–15
Seki M, Zyo K (1984) Anterior thalamic afferents from the mammillary body and the limbic cortex in the rat. J Comp Neurol 229:242–256
Shander D, Barraclough CA (1980) Role of the preoptic brain in the regulation of preovulatory gonadotropin surges in the hamster. Exp Brain Res 40:123–130

Sheridan PJ (1979) The nucleus interstitialis striae terminalis and the nucleus amygdaloideus medialis: prime targets for androgen in the rat forebrain. Endocrinology 104:130–136
Shimada M, Nakamura T (1973) Time of neuron origin in mouse hypothalamic nuclei. Exp Neurol 41:163–173
Shiosaka S, Tohyama M, Takagi H, Takahashi Y, Saitoh Y, Sakumoto T, Nakagawa H, Shimizu N (1980) Ascending and descending components of the medial forebrain bundle in the rat as demonstrated by the horseradish peroxidase-blue reaction. I. Forebrain and upper brain stem. Exp Brain Res 39:377–388
Sidman RL, Miale IL, Feder N (1959) Cell proliferation and migration in the primitive ependymal zone; an autoradiographic study of histogenesis in the nervous system. Exp Neurol 1:322–333
Sinding C, Robinson AG, Seif SM, Schmid PG (1980) Neurohypophyseal peptides in the developing rat fetus. Brain Res 195:177–186
Singer JJ (1968) Hypothalamic control of male and female sexual behavior in female rats. J Comp Physiol Psychol 66:738–742
Sofroniew MV (1983) Direct reciprocal connections between the bed nucleus of the stria terminalis and dorsomedial medulla oblongata: evidence from immunohistochemical detection of tracer proteins. J Comp Neurol 213:399–405
Sofroniew MV, Weindl A (1978) Projections from the parvocellular vasopressin- and neurophysin-containing neurons of the suprachiasmatic nucleus. Am J Anat 153:391–430
Spatz H (1925) Über die Entwicklungsgeschichte der basalen Ganglien des menschlichen Grosshirns. Anat Anz (Erg H) 60:54–58
Stephan FK, Zucker I (1972) Circadian rhythms in drinking behavior and locomotor activity of rats are eliminated by hypothalamic lesions. Proc Natl Acad Sci USA 69:1583–1586
Ströer WFH (1956) Studies on the diencephalon. I. The embryology of the diencephalon of the rat. J Comp Neurol 105:1–24
Stumpf WE (1968) Estradiol-concentrating neurons: topography in the hypothalamus by drymount autoradiography. Science 162:1001–1003
Stumpf WE, Sar M (1975) Hormone-architecture of the mouse brain with ^{3}H-estradiol. In: Stumpf WE, Grant LD (eds) Anatomical neuroendocrinology, Karger, Basel, pp 82–103
Swaab DF, Nijveldt F, Pool CW (1975) Distribution of oxytocin and vasopressin in the rat supraoptic and paraventricular nucleus. J Endocrinol 67:461–462
Swanson LW, Cowan WM (1975) Hippocampo-hypothalamic connections: origin in subicular cortex, not in Ammon's horn. Science 189:303–304
Swanson LW, Cowan WM (1977) An autoradiographic study of the organization of the efferent connections of the hippocampal formation in the rat. J Comp Neurol 172:49–84
Swanson LW, Cowan WM (1979) The connections of the spetal region in the rat. J Comp Neurol 186:621–656
Swanson LW, Cowan WM, Jones EG (1974) An autoradiographic study of the efferent connections of the ventral lateral geniculate nucleus in the albino rat and cat. J Comp Neurol 156:143–164
Swanson LW, Sawchenko PE (1983) Hypothalamic integration: organization of the paraventricular and supraoptic nuclei. Ann Rev Neurosci 6:269–324
Szabo J (1962) Topical distribution of the striatal efferents in the monkey. Exp Neurol 5:21–36
Szentágothai J (1964) The parvicellular neurosecretory system. In: Bargmann W, Schade JP (eds) Progress in brain research, vol 5. Elsevier, Amsterdam, pp 135–146
Szentágothai J, Flerko B, Mess B, Halász B (1968) Hypothalamic control of the anterior pituitary, 3rd edn. Akadémiai Kiadó, Budapest
Taber Pierce E, Foote WE, Hobson JA (1976) The efferent connection of the nucleus raphe dorsalis. Brain Res 107:137–144
Tejasen T, Everett JW (1967) Surgical analysis of the preopticotuberal pathway controlling ovulatory release of gonadotropins in the rat. Endocrinology 81:1387–1396
Ungerstedt U (1971) Sterotaxic mapping of the monoamine pathways in the rat brain. Acta Physiol Scand [Suppl] 367:1–48
van den Pol AN (1980) The hypothalamic suprachiasmatic nucleus of rat: intrinsic anatomy. J Comp Neurol 191:661–702
van den Pol AN, Cassidy JR (1982) The hypothalamic arcuate nucleus of rat; a quantitative Golgi analysis. J Comp Neurol 204:65–98

van der Kooy D, Carter DA (1981) The organization of efferent projections and striatal afferents of the entopeduncular nucleus and adjacent areas in the rat. Brain Res 211:15–36
van der Kooy D, Hattori T (1980) Single subthalamic neurons project to both the globus pallidus and substantia nigra in rat. J Comp Neurol 192:751–768
Vandesande F, Dierickx K, Demey J (1975) Identification of the vasopressin-neurophysin producing neurons of the rat suprachiasmatic nuclei. Cell Tissue Res 156:377–380
Vaughan E, Fisher AE (1962) Male sexual behavior induced by intracranial electrical stimulation. Science 137:758–760
Voneida TJ (1960) An experimental study of the course and destination of fibers arising in the head of the caudate nucleus in the cat and monkey. J Comp Neurol 115:75–87
Wasman M, Flynn JP (1962) Directed attack elicited from hypothalamus. Arch Neurol 6:220–227
Whitnall MH, Key S, Ben-Barak Y, Ozato K, Gainer H (1985) Neurophysin in the hypothalamo-neurohypophysial system: II. Immunocytochemical studies of the ontogeny of oxytocinergic and vasopressinergic neurons. J Neurosci 5:98–109
Whittier JR, Mettler RA (1949a) Studies on the subthalamus of the rhesus monkey. I. Anatomy and fiber connections of the subthalamic nucleus of Luys. J Comp Neurol 90:281–317
Whittier JR, Mettler RA (1949b) Studies on the subthalamus of the rhesus monkey. II. Hyperkinesia and other physiologic effects of subthalamic lesions, with special reference to the subthalamic nucleus of Luys. J Comp Neurol 90:319–372
Wiegand SJ, Price JL (1980) Cells of origin of the afferent fibers to the median eminence in the rat. J Comp Neurol 192:1–19
Wiegand SJ, Terasawa E, Bridson WE, Goy RW (1980) Effects of discrete lesions of preoptic and suprachiasmatic structures in the female rat. Neuroendocrinology 31:147–157
Witkin JW, Paden CM, Silverman AJ (1982) The luteinizing hormone-releasing hormone (LHRH) systems in the rat brain. Neuroendocrinology 35:429–438
Wolf G, Trautmann B (1977) Ontogeny of the hypothalamo-neurohypophysial system in rats. An immunohistochemical study. Endocrinologie 69:222–226
Wyss JM, Sripanidkulchai B (1985) An autoradiographic analysis of the time of origin of neurons in the hypothalamus of the cat. Devel Brain Res 21:89–98
Záborszky L, Makara GB (1979) Intrahypothalamic connections: an electron microscopic study in the rat. Exp Brain Res 34:201–215

Subject Index

Bold page numbers refer to illustrations

MIX
Papier aus verantwortungsvollen Quellen
Paper from responsible sources
FSC® C105338

If you have any concerns about our products,
you can contact us on
ProductSafety@springernature.com

In case Publisher is established outside the EU,
the EU authorized representative is:
Springer Nature Customer Service Center GmbH
Europaplatz 3, 69115 Heidelberg, Germany

Printed by Libri Plureos GmbH
in Hamburg, Germany